HEALTH
AND LIGHT

HEALTH AND LIGHT

The effects of natural
and artificial light on man
and other living things

by John N. Ott

Introduction by James W. Benfield, D.D.S.

ARIEL PRESS
Columbus, Ohio—Atlanta, Georgia

This book is made possible
by a gift from Jim and Nancy Grote
to the Publications Fund of Light

HEALTH AND LIGHT

Published by arrangement with
The Devin-Adair Company, Inc.

Printed in the United States of America by Ariel Press, P.O.
Box 1347, Alpharetta, GA 30201.

Library of Congress Catalog Card Number: 72-85334

ISBN 0-89804-098-1

To the men and women who believed in our early experiments and who continue to support our work at the Environmental Health and Light Research Institute.

INTRODUCTION

Although slow-motion photography has been known for many years, it has not been popularized until recently, when Americans have come to expect "instant replays" in sporting events on television. These are motion pictures taken at higher-than-normal speed. When they are then projected on the screen at normal speed they appear to slow the motion and make it possible to analyze a golf swing, determine the winner of a horse race, or follow a football player who receives a pass and runs for a touchdown.

There had been little use for the opposite type of photography, which gives the illusion of speeding up motion by means of taking single exposures at relatively long intervals, until John Ott began, while in high school 45 years ago, to experiment with what is now known as "time-lapse" photography. Fortunately for mankind, his hobby led eventually to a full-time career as a photobiologist. It is also fortunate that he had the fortitude to persevere against great odds; his chosen field was so new that much of the necessary equipment had to be designed by him and custom-built. Furthermore, some projects that he undertook required whole years to photograph even though the showing time of the resultant film was only a minute or two. Flowers and plants were among his first subjects. One of these films involved the growth of the banana from the emergence of the first shoot to the mature fruit. This project required ten cameras and two years to complete. Another sequence showing flowers, made to appear to dance by controlling light direction and temperature, took three years to produce—even though it lasted only two minutes on the screen.

Anyone who has observed individual cells under a microscope is aware of the fact that activity usually occurs so slowly that nothing seems to be happening. However, because of Ott's pioneering work in time-lapse photography, science has a new and invaluable tool which has almost limitless application. It is now possible, for example, to observe and record what happens within a single living cell—or to watch mitosis, or cell division, take place and to see changes that occur when a given stimulus such as a drug is introduced into the cell's environment.

It was while conducting a series of experiments in which individual cells were being photographed as certain drugs were introduced into their environment that Ott noted that changing the filters over the camera lens from one color (or wavelength) to another often had a greater effect on the cells than the drugs. This observation led to further studies on whole animals and the discovery that the quality of light is of great importance to both animals and man. It had long been recognized that the quality of light is important to plants, but Ott's work now showed that the process of photosynthesis in plants is only carried on at full efficiency in the presence of the complete spectrum of sunlight.

Man has lived on this earth for at least 100,000 generations and has been almost completely dependent upon the sun for light—until about five generations ago when Edison developed the incandescent lamp. Research has now demonstrated that the full spectrum of daylight is important to stimulate man's endocrine system properly and that he suffers side effects when forced to spend much of his time under artificial light sources that reproduce only a limited portion of the daylight spectrum. It therefore became obvious to Ott ten years ago that the design of artificial light sources should be changed to broaden their spectral analyses.

His attempts at that time to persuade two of the major manufacturers of light sources in America to do so failed, but it was my good fortune subsequently to be instrumental in prevailing upon the executives of a third company in the field to undertake such a project and to retain him as consultant. As a result, it has since produced a fluorescent light source that—for the first time in history—virtually duplicates daylight. Some remarkable testimonials have come from many industrial plants that have since installed this new lighting— such as substantial reductions in absenteeism and accident rates and marked increases in production.

It would not be presumptuous in the least to look at him as a twentieth-century Leeuwenhoek. As the 18th century Dutch scientist used the scientific "toy"—the microscope—and opened up new worlds to mankind, so has John Ott taken the motion picture camera, added Leeuwenhoek's "toy" and made a remarkable breakthrough in the study—and understanding—of light.

Recognition of his untiring research work has come to John Ott in the form of citations and awards from horticultural, scientific and medical societies, plus the Grand Honors Award of the National Eye Institute (in 1967) for an important contribution to eye care. In 1971, he was asked to give a seminar to scientists who were designing the specifications for the first United States space station. They wanted his counsel on the problem of growing vegetables for astronauts in space. His papers have been published in many scientific and educational journals, including those of the New York Academy of Sciences, the National Technical Conference of the Illuminating Engineering Society, the Fourth International Photobiology Congress at Oxford, the New York Academy of Dentistry and others.

There is much still to be learned about the effects of

light on plants, animals and man, but there is enough knowledge already available to provide important guidelines to manufacturers, architects and scientists who can directly influence the environment in which millions of people work and live. It has been my privilege to enjoy the opportunities of collaborating with John Ott in a small way for the past ten years. I firmly believe that the reader will gain important insights from *Health and Light.*

—James Winston Benfield, D.D.S.
October 1972

CONTENTS

PREFACE

Ever since the research of William Rowan in the '20s we have known that seasonal changes in the lengths of daylight and darkness have a significant effect on bird migration as well as upon mating periods for some species. Out of such studies, also, have grown the poultry industry's programs of lengthening short daylight hours in winter by means of artificial light in order to increase egg production. The response of the hens is due to the light energy entering the eyes and stimulating the pituitary gland. This has given rise to strong evidence that the endocrine system of mammals responds to particular wavelengths of *visible light* as well as other areas of the *total spectrum,* including the longer wavelengths of ultraviolet that penetrate the atmosphere.

This book is the outgrowth of extensive time-lapse photography, described in an earlier book, *My Ivory Cellar.* Some of that work will be summarized in order to provide the proper prelude to what we believe to be the pioneering studies of our Institute today. Actually, most of the research on the influence of light on the human endocrine system has grown from our observation of plant and animal growth responses to wavelength variations in the distribution of light energy—a result of time lapse pictures of plants growing and flowers blooming. This work has been developed over more than forty years.

As man has become more industrialized, living under an environment of artificial light, behind window glass and windshield, watching TV, looking through colored sunglasses, working in windowless buildings, the wavelength energy entering the eye has become greatly distorted from that of natural sunlight.

13

Much of the development of modern lighting has, unfortunately, been toward the use of light sources of increasing distortion. For example, the "natural white" fluorescent tube used in many hospitals to give the patients more color is greatly distorted from natural light. The sharp peak of energy in the red or longer wavelengths can make a pale, peaked patient look as though he had just come back from a vacation in a sunny climate. Flattering? Perhaps, but it creates an utterly false impression.

The tremendous significance of the rapidly developing body of knowledge about variations in wavelengths of light energy has finally spurred several big corporations to design products that permit the full spectrum of natural sunlight to enter the eye. Too little is known generally, however, about the importance of providing an environment of natural light indoors, where so many people must spend a major part of their time. It is our hope at the Environmental Health and Light Research Institute that this book will help chart new pathways toward that goal, as well as toward breakthrough findings in the fields of various ills that plague mankind.

—John N. Ott
Environmental Health and Light
Research Institute
Sarasota, Florida

Health and Light

One

THE LIGHT SIDE OF HEALTH

"You look in the pink of condition."

"I could tell that he was positively green with envy."

"When I heard her say that, I saw red."

Each of these phrases uses color in a figurative manner—relating a hue to an emotion, physical condition or attitude. When we use such expressions we recognize, of course, that we don't mean to have them taken literally. When was the last time you saw someone actually turn green with envy?

Still, as the scientific evidence comes in, we are becoming more and more aware of the fact that a very definite relationship exists between the colors that make up what we perceive as "white" or natural sunlight and our physical and mental health.

Perhaps the use of the word "colors" in the above sentence is a bit misleading. We generally think of a color as something we can see. But what *we* know as color makes up only a part of the spectrum.

For scientific purposes, different colors are defined in terms of a measuring system in which the wavelength is the standard unit. Each color has its own wavelength. The length of a color's wavelength determines its proper place in the spectrum.

But there are areas of the spectrum we cannot see. These areas are also measured in terms of wavelengths. The wavelengths which define these areas are either longer or shorter than those which define colors.

For instance, ultraviolet wavelengths are shorter than those to which the human eye is sensitive. Those beyond human perception at the opposite end of the visible spectrum are infrared. There are areas of the spectrum even shorter than the ultraviolet band or longer than the infrared band. Some are capable of penetrating through most ordinary types of building materials as easily as the light we see passing through glass.

Wavelengths shorter than ultraviolet and longer than infrared are usually referred to as radiant energy. It is possible to be in an environment in which the eye cannot perceive anything except total darkness and yet be exposed to radiant energy in one form or another with the organism responding accordingly. The latter statement is one which poses a fascinating mystery— a mystery I became more and more compelled to probe into. And so I found myself becoming more deeply involved with responses to light and radiation as I pursued my work in time-lapse photography.

I began to open up whole new areas of investigation into light, particularly when I added the microscope to my equipment. Microphotography has been known and used for many years; I put it into motion via time-lapse and found things no one had ever seen or suspected.

I was able to observe the movements of cells in Elodea grass, but more than that, in the course of my own experiments I learned that the cells behaved quite differently under different colored lights. Mostly, these cells perform in an established pattern when exposed to any natural sunlight condition. I soon found, however, that they broke the established pattern and displayed many variations when different filters were used in the microscope light. I could make the cells go in different directions; I could cause some of them to stand still while others took up the new patterns.

While plants were the first living things I worked with, my quest soon took me to cells from animals. Here again I found that I could create radical changes within the cells by changing color in the microscope. I could increase their metabolic activity; I could kill them.

Working with live animals—laboratory mice—I discovered that various kinds of lighting conditions could affect them physically. Not only did the changing lights cause external physical changes: the lights had a definite effect on their sex lives and life spans.

In order to cover all the areas that presented themselves as worthy of investigation, I found myself entering into worlds other than my own, worlds which ranged from Hollywood motion pictures to TV stations, mink farms, a prison, a restaurant where all the employees seemed inordinately healthy, and a dozen other areas I might not have entered normally.

My primary work, of course, centered on my own time lapse studio and in various scientific laboratories where I was privileged to work or observe.

I was to find clues in very unlikely places—clues to the effects of light on life—and health. Yes, it became clearer after a while that there was some mysterious link between light and the mental and physical health of humans.

The key to all this seemed, to me, to lie in the simple act of light entering the human eye. I found many dramatic examples of changes in health when sunglasses were worn by an individual—or taken away. I found interesting improvements in physical conditions when the subjects were exposed to full-spectrum lighting or exposed to natural sunlight over long periods—without benefit of ordinary glasses.

I learned, with total fascination, of the Congolese who decided to wear sunglasses as status symbols, and what happened to their general health collaterally—or coincidentally.

Most of the things I learned were fascinating and compelling, but one instance, stumbled upon acciden-

19

tally and which led to some of the most dramatic experiments I ever performed, was frightening in its implications. The results of these experiments caused congressional investigation, recall of a product by a major American manufacturer, unplanned and hurried testing in industrial laboratories and radical revision of certain standards in the product involved. The story, described in detail later on, might be called "The Case of the Tired Children."

Out of all these quests and probings have come some good things, but I still remember the frustration of one who presents a case to science and finds the academic backs turned on him to a large degree. I realize that much more work must be done in the field of light, particularly in those areas where I have shown dramatically that light can and does affect human health; the problem seems to be in getting enough recognition and agreement from science itself to undertake this additional and extended work.

Fortunately, some of the good things I mentioned above have already made progress. One of the ideas I have stressed in linking light with health lies in the fact that ordinary eyeglasses, windows in homes and automobile windshields screen from the eyes most of the ultraviolet which reaches us in natural sunlight. And depriving the human of that ultraviolet can become a strong obstacle to improving health. We humans manage to survive even with that deprivation, but now industry is taking some steps to restore the opportunity of receiving ultraviolet in the way in which it should be introduced into the system. Several companies are now making new types of eyeglasses and contact lenses which will not block ultraviolet. These neutral gray sunglasses and tinted contact lenses are designed to cut down all wavelengths evenly, including the ultraviolet, so that the natural balance of light will not be upset and colors distorted.

New types of fluorescent lights which closely duplicate natural sunlight are now being made. What this does, simply, is bring a close approximation of natural sunlight indoors.

These are but first small steps in the journey toward understanding and proper utilization of light. And they're only surface-scratchers. One of the areas I am most vitally interested in is cancer research and, as you read along, you'll learn of some of the startling results I have gained by linking light to cancer therapeutically. And therein lies another of my major frustrations, but that's a story that unfolds later in this book.

As we have seen, the kinds of intensities of light we are exposed to have a great deal to do with our health. Who knows? Perhaps sometime in the near future relationships between the full spectrum of the sun's natural rays and health will be better understood. Then, to keep well and happy, we may find ourselves being put on "light diets" in the same way we go on food diets today.

Yes, there is more to the rainbow than meets the eye.

Two

HOW IT BEGAN

The principle of time-lapse photography is very simple. It is just the opposite of slow-motion pictures, with which most people may be more familiar. Instead of slowing fast motion down, it speeds up many times faster than normal such subjects as a flower opening, or the complete growth of a plant on the screen in a few seconds. The actual time represented may have been several months or even years. It is somewhat similar in principle to the animated cartoon type of picture. However, instead of drawing each frame or individual picture to be photographed by hand with the action advanced a little each time, live growing subjects are used. It is then necessary to wait for a little growth to take place between each picture. In some instances, with rapidly developing microscopic subjects, this may be only a few seconds. With the opening of the petals of an average flower, it would be about every five or ten minutes, and with something like the development and ripening process of fruit, that takes a much longer period of time, it might be one picture every hour or two. The time interval between pictures may have to be changed as the plant goes through different stages of development, such as the opening of a blossom, to the slower maturing process of the fruit. Nevertheless, the individual pictures must be taken regularly, day and night.

In observing the growth of plants, I noticed that the flowers and leaves always faced into the light and that the leaves noticeably drooped from lack of water, but would quickly revive when given a drink. I always placed the camera so that the constant daylight came from behind and thus the flowers in facing the daylight would also be facing directly into the camera.

One night I dreamed up a wild idea of controlling the light, temperature and moisture to make the leaves of the plants move in different directions. To accomplish this, it was necessary to construct special flower pots that would move around on wheels. In each pot was placed a small electric heating element and a water tube. Many different flowering plants were tested and primroses were found to respond best to this treatment. A few more refinements on my timing contraption and everything was set. The flower pots were pulled around on a track like an old fashioned cable car, but at a speed of about 1/2 inch an hour. The heating elements were turned on at the proper time to wilt the leaves down and the plants were given just the proper amount of water to revive them again. A battery of lights was first turned on one side and then the other, which would attract the leaves from side to side. Thus it was possible to move the plants around a miniature stage and control both the up and down motion and also sideward motion of the leaves. It only remained to synchronize this motion to music. This sequence lasted only two minutes on the screen but required five years to complete, including an interruption of two years while I was in the Navy. These dancing primroses have always caused considerable comment, and later I used them as the opening theme for my television programs.

Obtaining some of the necessary electrical equipment for the dancing primroses was often a problem, especially following the war period while priorities were still in force. Everybody always wanted to know why I needed a particular type of switch and what I intended doing with it. At first I tried to avoid any direct reference to the waltzing primroses but was always given some excuse for a further delay in delivery of the equipment needed. Finally I told the manager of the sales department of a particular company that I had to have a certain automatic switch in order to take pictures of my dancing primroses doing a Strauss waltz. It was hard for me to keep a straight face. The sales manager must have thought the easiest and quietest way to get rid of me was to let me have the switch.

The waltzing flowers were not the only subjects that required special equipment. I had experienced considerable difficulty in trying to make time-lapse pictures of corn growing in the glass greenhouse. Although the pictures were good, the corn always grew spindly. The ears would not develop to normal size, and the leaves were long and narrow. It was not possible to take time-lapse pictures of corn growing outdoors because of the wind and weather. The leaves would be in a different position for each picture. Finally I tried growing corn outdoors and letting different plants develop to different stages. Then I built a make-shift enclosure around them and tried to photograph the formation of the ears and tassels after the plant had grown out in the open and up to the time these parts began to appear. In making the enclosure, it was more convenient to use some of the new plastic sheeting that had just come on the market. The corn grew much more normally under plastic than it did under glass so I began experimenting with the growing of corn and other plants under different kinds of plastic.

Most ordinary glass stops over 99 per cent of the ultraviolet radiation whereas some plastics allow approximately 95 per cent or more to pass through. The practice of old time experienced nurserymen in completely removing the glass sash from the cold frames during the daytime to expose the young seedling plants to direct sunlight always interested me. The improved growth warranted the additional labor of completely removing the sash in place of raising it a little for ventilation during the daytime, and replacing it again at night when there might be danger of frost. The results of using plastic in place of glass were so much better that I decided to build a new plastic greenhouse entirely without glass. This was unheard of at the time, so I had to call on my friend, Herman Schubert, to fabricate the entire structure in his machine shop. It was built in sections, then dismantled and moved to the selected location in the backyard where it was reassembled again.

The roof, east, south and west sides were made of clear plastic, so the growing plants would receive much more sunlight. These large areas had to be covered with automatic shutters that closed each time a picture was taken in order to have the same amount of photographic light as in the basement studio. These large shutters, like giant Venetian blinds, worked independently of each other and automatically followed the sun so that the louvers all remained parallel to the rays of light as it moved across the sky. In this way they created the least amount of shadow and let in the maximum amount of direct sunlight. The north wall was solid and was painted sky blue to act as a photographic background. Sky blue seemed to be the most natural color for a background and made it easier to match backgrounds of time-lapse pictures taken in the greenhouse with regular shots made out in the field. The new greenhouse also had more head room so tall corn and small trees could be grown inside. This made it possible to start taking time-lapse pictures of many more new subjects, and also created many more new problems.

When the new plastic greenhouse was completed to the point where time-lapse cameras could be started, the Ferry-Morse Seed Company wanted a time-lapse picture of one of their varieties of morning glories. This certainly appeared to be one of the simplest assignments I had ever undertaken. I planted some seeds in another small lean-to glass greenhouse by the garage. This was a good place to start some subjects and grow them until they were ready to photograph. As the morning glory vines neared the budding stage, I moved them from the glass house to the plastic house. Everything went well until the buds were just ready to open. Then bud after bud would collapse. Could it be the result of having moved or changed the growing conditions during the bud development period? Any such slight change of conditions should have been all for the better. Nevertheless, I started more seeds right in the plastic greenhouse where they could be photographed and nothing would have to be moved or changed in the slightest during the entire growing period. Again, exactly as before, the buds collapsed just when they should have opened. I could remember having seen morning glories growing in glass greenhouses with flowers in full bloom, so everything was cleared out of the glass greenhouse. Then cameras, lights, timers, shutters—the whole works—were moved in again, and more morning glories were started from seed. This all took time and summer was just about over, but with any luck there should still be time for one more crop of morning glories. Though still disappointed about the results in the plastic greenhouse, the morning glories and cameras were all installed in the old glass lean-to alongside the garage. By slightly swallowing my pride— and with a little luck—this picture might still be completed during the current year. But, as the buds again reached the point when they should open, the same thing happened—no luck.

The fact that the buds still refused to open in the glass greenhouse indicated perhaps the problem was not with the plastic after all. This in itself was quite gratifying even though I still did not have the needed pictures or any real clue to why the morning glory buds persisted in collapsing without opening. One thing learned right at the start in trying to take time-lapse pictures was not to photograph plants out of their normal growing season. It is hard enough with most plants, if not practically impossible, to grow them at all out of season, let alone trying to get flowers worth photographing. People are not interested in looking at a poor picture, even though it may be of the impossible. This project had to go down in the records as unfinished.

But I was to get further clues to the solution of the morning glory problem from a surprising source. During the very early days of television, I was asked to appear on several programs and show time-lapse pictures of flowers. These unplanned appearances, on one of Chicago's first stations, went so well that I was invited back again and again. After only a few Sundays of help- ing the station's program director fill a half-hour, I was surprised to see myself listed in the paper as a TV personality. The Sunday half-hours continued, and even expanded to include my participation either live or on film in a number of established programs including "The Home Show," "Today," "Out On The Farm," and "Disneyland."

Late that fall when the morning glories were causing so much trouble, one program was on the subject of chrysanthemums. In researching my subject, I was fascinated to learn how they are grown out of season, most varieties now being available every month of the year. Largely, chrysanthemums set their buds toward late summer as the daylight hours shorten. It has now been proven that some plants, such as wheat, require a certain number of hours of sunlight before the plants will head up, but with chrysanthemums it is just the

27

reverse. The plants of course need sunlight to grow, but it is the lengthening of the dark or night period that controls the setting of their flower buds. When the length of the dark period amounts to approximately ten and a half hours or more, the buds begin to develop. It is now common practice for commercial growers to cover their chrysanthemum plants with black cloth suspended on wires or iron pipe frames. During the long daylight hours of the spring months, they are covered over about four-thirty in the afternoon, and not uncovered until about eight-thirty the next morning. The effect is to lengthen the night period prematurely or artificially and the chrysanthemums set their buds and come into flower ahead of their normal blooming period. Likewise, if ordinary electric lights are turned on in the evening as the days grow shorter, and the night period held under ten and one half hours, the plants will keep on growing taller, and delay setting their buds.

This sounded like a possible clue to my morning glory trouble. Maybe the photographic lights going on every few minutes for five seconds all night long were interrupting the normal dark period. My morning glories had no problem of setting buds, they simply wouldn't open. The light only affected the setting of buds on the chrysanthemums. After they were once set and showing a little color, they would continue to develop regardless of the length of the day or night period.

I was also trying to take time-lapse pictures of poinsettias for my Christmas TV program, and ran into similar problems. The poinsettia flowers—that is, the colorful bracts—literally stopped in their tracks as soon as I started photographing them. It made no difference how far along the flower was. I heard a story from Honolulu that the Chamber of Commerce there

28

had run into the same difficulty when an attempt was made to turn floodlights on some of the poinsettias in one of the parks. It is also known that a street light, too near a greenhouse of poinsettias, will cause them all kinds of blooming problems. Commercial poinsettia growers have quite recently started to control the light and dark period to bring their plants into peak bloom on Christmas Day. Poinsettias are another flower that sets buds and comes into bloom as the dark night hours grow longer during the fall and winter period. Accordingly, as the length of daylight shortens earlier in the season the farther north you go, poinsettias also bloom earlier up north.

Now it is customary to turn the lights on in the green houses for approximately two hours at midnight for a ten-day period commencing September twenty-ninth. That is when poinsettias ordinarily begin to set their buds in the latitude of Chicago. By turning the lights on and interrupting the night period, the process of setting buds is delayed, and now the height of the blooming period is attained right on Christmas Day. In this way the plants are much fresher and will last longer.

All this seemed to tie into my morning glory problem, but just *how* was the real question. Nothing seemed to fit exactly, but nothing further could be done with morning glories anyway until spring. The winter went quickly, and soon it was time to plant more seed. When the first buds were ready to open, I had morning glories all over the place—in the plastic greenhouse, in the basement studio under the big skylight, in the glass lean-to greenhouse and, this time, outdoors on the garden fence. The morning glories outdoors bloomed fine, but the ones indoors still insisted on collapsing just when they should open. I tried cutting down the intensity of the photographic lights to the point where pictures could no longer be taken, and still they collapsed. Then I got an idea and, accidentally, something happened at the same time that gave me the answer.

The name morning glory is slightly misleading. I had assumed that they open to their full glory with the morning sun as it rises. But one morning when I was up well before sunrise, I noticed that the flowers outdoors were already open. In other words, morning glories are a night blooming flower. Interesting, but still, so are iris and others that presented no problem photographing. The night-blooming cereus readily opens only at night even though photographic or other bright electric lights are turned on in the same room. I decided to stop taking pictures in the glass greenhouse altogether. That night the morning glories opened normally. The next night I hung an extension cord with a light on the garden fence where the morning glories had been blooming normally. It was connected so that it would flash on each time a picture was taken in the plastic greenhouse. The next morning the outside morning glory buds were all collapsed within a perfect circle around the electric light. This showed definitely the problem was clearly due to the photographic lights interrupting the night period.

Another incident at this same time came as a result of concern with the budget. Expenses on this subject had to be charged to the research and experience account, so I was quite conscious of piling up additional film costs. I happened to have a short end of daylight type color film which I thought might as well be used up. Ordinarily I used commercial Kodachrome, which is color-balanced to regular photo-flood lamps. As all photographers who take any color pictures know, daylight film is balanced to sunlight, which is somewhat bluer than ordinary artificial lights. Indoors it is necessary to use either a blue filter over the lens with daylight type Kodachrome film or special bluish lights.

This time I used blue lights instead of the regular ones without giving it a second thought. The next morning when checking the morning glories, I really did give it a lot of thought, as several of the buds had opened halfway for the first time. What was it? The only possible difference was the use of blue photo-floods to go with the daylight film. Actually, several of them had been used, and the bluish light was considerably brighter than the regular lights that wouldn't work at all. Could it be because the light was a little blue? Maybe wavelengths of light had something to do with these morning glory buds.

That night I put additional blue filters over the slightly blue photo-floods. The next morning the morning glories were open full and perfectly. The pictures were, of course, too blue, but after a little experimenting with red filters I could get a pretty well-balanced color picture. The blue filters, in effect, were filtering out the red wavelengths from the photographic light. It was the red wavelengths of the spectrum in the photographic lights interrupting the normal dark period at night that was the controlling factor in preventing the morning glory buds from opening.

Another equally troublesome subject was in the works at the same time. This was a time-lapse picture of the growth of a pumpkin for Walt Disney's *"Secrets of Life."* This project presented various new problems, particularly in regard to the sex life of a pumpkin. The pumpkin is a member of the cucumber family, which includes most of the squashes and melons, and is known as *monœcious.* This means that it has separate male and female flowers on the same plant.

To start, I planted a pumpkin under the skylight in the basement studio. Next, I hung some fluorescent lights over it in order to supplement the natural daylight and approximate full summer sunlight intensity.

The pumpkin vine grew well, but all the female or pistillate pumpkin-producing flowers turned brown and dropped off very soon after they formed. The staminate or male pollen producing flowers grew vigorously but were of little use all by themselves. Needless to say, no pumpkins developed so I had to try it again the next year. Meanwhile the fluorescent light tubes had burned out, and had to be replaced. The hardware store was out of the regular fluorescent tubes used the previous year and rather than wait for them, I used some daylight-white tubes. These were generally less desirable and in less demand, as their slightly bluish color made ladies' lipstick look a ghastly purple.

I planted more pumpkin seeds the second year under the new fluorescent lights and watched carefully. The vines seemed to grow just as well. The flower buds began forming, but now all the male flowers turned brown and dropped off, and the female flowers developed vigorously. Thus for the second time after having photographed a vine every five minutes over a period of months from the time the seed was planted until it had grown over fifteen feet, I was confronted with the problem of all flowers of only one sex. This time, I had a perfect female or pumpkin-producing flower which I figured would open the next day, and nothing but pictures of the male flower—and a year old at that.

The next day the flower was open and would have to be pollinated before closing that night if it was to produce a pumpkin. In this delicate situation I decided to call my friend, Dr. Harold Tukey, head of the Department of Horticulture at Michigan State University. I asked if he had any pumpkins growing at this time of year in the department's experimental greenhouses, and also if it might be possible to use any kind of artificial hormones. Dr. Tukey had no pumpkins and advised that real pumpkin pollen would be necessary under these circumstances.

Next I called Dr. Dorsey at the University of Illinois, but he had no pumpkin pollen at the moment, either. Then I called Dr. Julian Miller at Louisiana State University. Possibly the pumpkins might be blooming in Louisiana, but Dr. Miller told me there would be none for a week or ten days. This would be much too late for my beautiful pumpkin flower which was now in full bloom.

Maybe the pumpkins might be in bloom a little farther south, so I called Mr. Deatrick of the Flagler Hydroponic Gardens in Miami. After explaining the urgency of the situation, he said he would see what he could do. He called me back a little later in the morning, saying that he had spread the word. By noon an urgent appeal for pumpkin pollen had been published in the early edition of the Miami paper and broadcast on the media so my beautiful little lady flower wouldn't die a spinster. By two o'clock that afternoon, someone living in Miami called in that she had a pumpkin vine with male flowers in bloom and offered it in this emergency.

Next, the phone rang. It was a vice president of Eastern Air Lines. He had heard of the plight of my lady pumpkin flower in full bloom and offered their facilities in the emergency. He arranged to have the whole pumpkin vine dug up and placed aboard a non-stop plane to Chicago. I dashed out to the airport and waited for the plane. Newspaper photographers and reporters had gathered, but no one really knew who the important celebrity was. It obviously had to be a movie star or foreign royalty. No one would believe me when I told them it was King Pumpkin. The passengers were all held on the plane until the pumpkin was unloaded and delivered to me. I rushed it out to my time-lapse studio, where I introduced it to the lady in waiting. Now

I am being called Uncle John by a lot of little pumpkins and particularly the one that starred in Walt Disney's *"Secrets of Life."*

The fact that either male or female flowers can be brought forth by controlling slight variations in color or wavelength of light opens up some interesting possibilities for investigation. The next step will be to shade various parts of the plant from the fluorescent light, and try to discover whether the control comes through the leaves or the flower itself, and just what part of the flower at what stage of development—and how long the exposure to light must be.

In any case, I had solved the morning glory problem, even though the pumpkins still left me a little in the dark.

Three

THE ELECTROMAGNETIC SPECTRUM

From the preceding, it's obvious that light exerts a profound effect on plants and, as will be seen later, on all animal life as well. It would be a good idea to pause here for a brief look at some charts which show the composition of the total electromagnetic spectrum. These typical spectral charts explain its various parts and show the differences in the wavelength energy distribution between various light sources.

The chart of the total electromagnetic spectrum shows the continuity of the different wavelengths pictorially rather than to scale. The length of the different wavelengths indicated on the chart is from 0.0000000000003937 of an inch for cosmic rays up to a length of 545 meters for the longest radio broadcast wavelengths and 3,100 miles for electric waves produced by a 60 cycle generator. As can be seen in the chart, the wavelengths of visible light represent only a very narrow band near the center of the total electromagnetic spectrum.

Sunlight is a broad, continuous spectrum peaking a little in the blue-green. It then cuts off abruptly in the ultraviolet at about 2900 angstroms* because of the filtering effect of the earth's atmosphere. The longer wavelengths of ultraviolet that do penetrate the atmosphere at intensities comparable to visible light are sometimes referred to as near ultraviolet, and include the so-called black light ultraviolet. The shorter wave-

* The angstrom is a unit of length, one ten billionth of a meter in diameter, used in optical or spectral studies. It is applied to macroscopic measurements such as thickness of light wavelengths, liquid films, molecular diameters, etc.

ELECTROMAGNETIC SPECTRUM

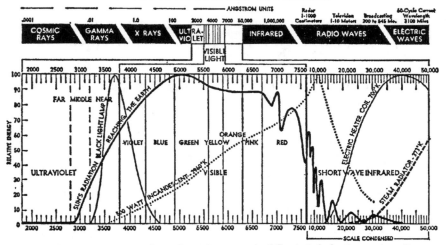

The human eye sees less than 1 percent of the total electromagnetic spectrum. Little is known about the mysterious light sources at either end of the visible spectrum—the ultraviolet, infrared, and so-called background radiation—but evidence now seems to indicate that they exert a profound influence on the physical and mental health of animals, plants, and man.

lengths of ultraviolet that do not penetrate the atmosphere are sometimes referred to as far ultraviolet, and include the germicidal wavelengths that can be very harmful.

The ordinary incandescent light contains virtually no ultraviolet, is lacking in the blue end of the spectrum, and produces its maximum energy in the infrared wavelengths that are not visible to the human eye, but do produce heat. This is why the incandescent light bulb gets so hot and is less efficient than the cooler burning fluorescent light tube. A tungsten filament operates at twice the temperature of molten steel, and is hot enough to melt asbestos or fire brick. The fluorescent light operates on quite a different principle from that of the incandescent bulb. It is filled with argon gas and mercury vapor. At each end of the fluorescent tube is a cathode. When electric current is turned on, the cathodes discharge electrons and a flow of current takes place through the mercury vapor and produces an electrical arc. This arc is an efficient producer of short wave ultraviolet light concentrated at one particular wavelength of 2537 angstroms. This is called a mercury vapor line. There are also other mercury vapor lines of lesser intensities in both the ultraviolet and visible wavelengths, as shown in the accompanying spectral charts of fluorescent tubes in common usage.

The 2537A short wave ultraviolet line causes the phosphor coating inside the glass tube to fluoresce, thus converting the invisible short wave ultraviolet to longer wavelengths of visible light. Different phosphor materials fluoresce at different wavelengths, or colors, and the proper blending of different phosphors produces the different types of fluorescent lights, such as cool white, warm white, daylight white and some of the deeper colors as well. The type of glass used in the tube of a fluorescent light will allow the longer wavelengths produced by the phosphors to penetrate, but filters out the short wave ultraviolet produced in the mercury arc, just as the atmosphere filters the short wave ultraviolet from sunlight. A short wave ultraviolet, or germicidal, fluorescent tube produces the same basic mercury arc. However, the tube is made of a type of glass that will permit the shorter ultraviolet wavelengths to penetrate, and it is not coated with any phosphors, so that the light source is basically the 2537 angstrom wavelength of the far ultraviolet part of the spectrum which will kill bacteria and can be very dangerous and harmful to humans.

Photobiological responses to specific colors, or relatively narrow bands of wavelengths within not only the visible spectrum but also the ultraviolet, give further evidence of the need for scientific control of experimental laboratory light sources. If a certain photo-receptor mechanism responds only within the range of near ultraviolet, then it becomes rather meaningless to study its responses to high intensity light sources containing no ultraviolet. If the responses are to either red or blue, then cool white fluorescent tubes would not be the best light source, because their peak of energy is in the yellow-orange part of the spectrum. Consideration should also be given to what the wavelength resonance is of any drugs, vitamins or other components of the diet being studied to determine if they coincide with any of the mercury vapor lines which in turn vary in intensity in different types of fluorescent tubes. Mercury vapor lines are conventionally represented at 100 times their true width and at only 1/100 of their true intensity, thus giving one a distorted impression of the actual values involved.

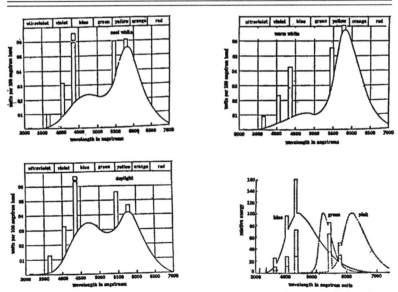

Spectral wavelength charts—cool white, warm white, daylight white and various colors of fluorescent tubes. (Source: Sylvania.)

Four

RELUCTANT APPLES
AND TIMID TIGER LILIES

The same year my pumpkin vine was producing all male flowers, I was also having problems with an apple that refused to ripen. Walt Disney wanted to include a picture of the growth of an apple in the same film, *"Secrets of Life."* It was hardly practical to move an apple tree down into the basement, so I built a complete time lapse studio in miniature on a scaffold by the apple tree in the front yard. It consisted largely of a glass window or skylight in the top of a large box, equipped with shutters that would close to keep the sunlight out momentarily each time an individual frame was exposed on the moving picture film. It also contained the necessary timing equipment to operate three cameras, the overhead shutters, and photographic lights. Two automatic thermostats controlled an electric heater and ventilating fan to maintain the proper temperature in the box and prevent over-heating in the direct sunlight.

A branch of the apple tree that had the best looking buds was selected, and the large box-like time-lapse studio was placed around it. Both the subject and equipment were then completely protected from wind and rain. The apple branch was fastened securely so it would not move during the time required for the dormant bud to develop into a nice juicy red apple. The entire tree had to be battened down with many wires and turnbuckles to hold it rigid and motionless during a severe thunder or windstorm.

Everything was completed and ready to go about the middle of March. The switch was turned on and the project officially started. If all went well, this picture would be completed by apple harvest time in October. The cameras had to be checked at least once each day and a careful watch maintained for any insects or disease that might harm the apple. All went well for a while. The buds opened on schedule and were large and healthy looking. Pollen from several other varieties of apple trees was collected and a small camel hair brush used to hand pollinate the blossoms being photographed. Ordinarily this is done by honeybees, and frequently commercial orchard growers hire beekeepers to bring their hives into the orchards during the blossom period.

There was no problem in hand-pollinating the blossoms in the box, but I needed a close-up of a bee itself on a blossom, and this was not easy. The bees never stayed on any blossom long enough to set up a camera and focus it properly. The blossom also had to remain perfectly motionless, as the field and depth of focus on such an extreme close-up were very critical. Finally I fastened a twig with a freshly opened apple blossom on one end of a board with a sky-blue background. At the other end I mounted my camera and waited for the bees from a "planted" beehive. However, they completely ignored both me and the blossoms. I kept poking it a little closer to the hive opening where the bees were streaming in and out. I moved it around and wiggled it to attract their attention. Suddenly, as though someone had given a signal or command, the bees all came at me like dive bombers. They got in my hair and buzzed and swarmed all over me, but surprisingly enough, not one stung me. I got the idea, though, that my presence was not appreciated near their hive and quickly retreated. Then I noticed bees in a tree in another yard nearby. When I placed the end of the board with the apple blossom on it up in this tree, the bees would

40

accept it, and I was able to get a good close-up of a bee at work collecting nectar and pollen.

Soon the blossoms were dropping, and the small apples were beginning to take shape. As the pictures later showed, apples grow during the daytime and relax at night. The effect on the screen was like someone blowing up a balloon a little at a time. During the entire summer I continued to watch the development of the several apples on the branch inside the time-lapse box and compare them with the other apples on the same tree that were out in the open. Everything seemed to be going along perfectly normally until all the apples not in the time-lapse box began to mature and turn a nice red color. The apples inside the box were still green and continued to grow larger and larger. The increased size was fine, but Disney wanted the picture to show the apple turning red. Spraying the apples with various chemical products that were supposed to make fruit develop better color had no effect. At last the apples outside my box began dropping off the tree. Inside the box they kept on growing until the weather was so cold that they froze solid—and still a disgustingly healthy-looking green color.

This was another disappointment and also a very important subject. There was no real clue as to what could have been the trouble. Down came all the equipment, and down came the unsightly box from the apple tree by the front door of the house. I thought about this all winter and discussed it with many friends and experts on growing apples. The best thing to do was to try it again the next year on a different variety of apple. To make doubly sure of getting a picture on the second try, two scaffolds were built by two other varieties of apple trees. On each one went a big box with all the equipment. I watched and waited. The blossoms opened and were hand-pollinated again. The small apples

41

began to take shape. Day by day, all summer long, I waited and watched them grow larger. If you think it takes a long time for a kettle to boil while you're staring at it, try watching apples grow for two years.

In the past no difficulty had been encountered with the ripening process or coloration in making time-lapse pictures of many other subjects including peppers, tomatoes, and various fruits and vegetables from bud to full maturity. (Tomatoes will turn red even when picked green and stored in complete darkness.)

As the season progressed and the time of year rolled around again, I watched faint traces of red color begin to show in all the apples in all the trees except those being photographed in my two time-lapse boxes. There the green color persisted, and again they continued to grow larger and larger. I double-checked the temperature controls in both boxes and found only approximately two degrees variation from outside. Taking into consideration the wide fluctuation of temperatures from day to day and variations between daytime and nighttime, this slight difference certainly could not be enough to matter. What could be preventing the apples in the box from maturing?

In desperation, I removed the glass from the window over the apples and replaced it with the new plastic material that let more of all the sun's rays penetrate and particularly the ultraviolet and shorter wavelengths. These are the ones that ordinary glass will not transmit. Within two days the apples in the boxes were showing a nice red color. The picture was completed just in time to be included in Walt Disney's film, *"Secrets of Life."* I was convinced that the maturing and ripening process of an apple can be prevented by filtering certain wavelengths of energy from normal sunlight.

For quite a few years I used the waltzing primroses

as the grand finale to my lecture film. Everyone seemed to enjoy it, and as many times as I had seen the flowers dance around and take a bow, they never became tiresome to watch. Five years was a long time to spend on one short sequence, and I felt it should last a long time. As new pictures were completed, I would replace some of the old ones, but the waltzing flowers were like a trademark, and to leave them out was unthinkable.

Times change, though, and one day two of my sons tactfully tried to explain to me that the waltzing flowers had seen their best days and were getting out of date. They thought I would do much better to introduce some rock and roll or dixieland rhythm. It was hard to visualize primroses doing rock and roll, but gradually the idea sank in. Maybe some other kind of flowers would respond faster, tiger lilies, perhaps. It required two or three days for a primrose leaf to wilt down and then revive, but many flowers would open and close their petals in response to light and heat alone, and much faster than primroses wilting. The accepted theory was that the petals of flowers opened and closed from a more or less mechanical response to light and heat. If the timers on the skylight controls were adjusted, it would be possible to shorten the day and night periods artificially to just a few hours. The individual pictures could be taken at shorter intervals, and perhaps several days' work could be accomplished in one ordinary 24-hour period. The day and night controls on the thermostat could also be adjusted so that the temperature would correspond to the shorter light and dark periods. The air-conditioning equipment would cool the studio down 10 degrees while the skylight shutters remained closed to simulate the cooler night air. There was no doubt about it now, the dancing flowers were going to keep up with the times. I would make the petals open and close three or four times during one ordinary day.

The music, supplied by a troupe led by a neighbor's son, was recorded on magnetic tape, and then re-recorded onto an optical film track so the various vibrations could be analyzed. The motion could then be plotted for the tiger lilies' growth. A sound track can be pulled through a sound reader so that the sounds of different instruments can be heard and marked on the film. However, after studying the sound track for a while, it is not too difficult to read the characteristic vibrations visually and pick up the beat of the rhythm as well as the different instruments. My procedure was to number each individual frame on the picture part of the film and then sketch the position the flowers had to be in at that particular point. Next, the length of time necessary for the plants and flowers to reach that state of development had to be estimated. Then the number of frames along the sound track between the various points of growth development were counted. A little simple arithmetic, and the automatic timers were set. All that remained to be done was to grow the plants so the flower buds would reach the predetermined stages of development on schedule.

Certain flowers like roses, peonies and tulips open their petals during the daytime and close them at night. Many other flowers open only once and stay open until they fade away. Another group open during the night-time and close their petals during the daytime. Then there are those that open for the first time during the night and stay open. Four O'Clocks open at the end of the day along towards dusk, and tiger lilies are in the group that open at the end of the dark night period as the eastern horizon begins to show the first faint glimmers of light. The opening and closing habits of different flowers are one more factor to contend with when adjusting the timing mechanisms that control the cameras, lights, shutters, temperature and automatic watering devices used in making flowers dance in time to music.

At last came the day when the flowers would start dancing to dixieland rhythm. When the pictures were finished and projected on the screen, the action had to coincide exactly with the rhythm of the music. While the pictures are being taken, the flowers of course do not hear any music, as this has already been recorded. The speeded-up action necessitated by dixieland rhythm means that one single movement of the petals must be accomplished in two hours now instead of twelve as in the past. This faster action still cannot be seen at normal speed as the flowers grow, but through time-lapse photography, the flowers when projected on the moving picture screen move about in rhythm with the music.

Again, everything worked as planned at the outset. The skylight shutters opened at sunrise, and the heat came on as scheduled. The petals on the flowers all opened up perfectly. With the increased rate of taking individual exposures, the normal full day's work was accomplished by 10 o'clock in the morning. As planned, the skylight shutters closed, the heat went off, and the air conditioning started. All the flowers closed their petals just as they would normally do during the night. At the continued increased rate of picture taking, the night's work was completed by 12 o'clock noon. Automatically, the skylight shutters opened again, the air conditioning shut off, and the heat turned on. The sun was bright and clear and at its maximum intensity. The closed flowers were in full direct sunlight. This would most likely make them open a little faster than the early morning sun, so I made a slight change in the timing schedule, as this was one thing that had been completely overlooked. I watched closely and waited. Ten minutes went by, and nothing happened. Fifteen minutes, a half hour, and still the petals remained in their closed position. I changed the timing mechanism from the faster schedule to an extremely slow one, but by now everything was completely off schedule. The action of the flowers would already be out of synchronization with the prerecorded music. This group of flower subjects was spoiled, but perhaps I could use them to experiment with and make them open with additional artificial light and higher room temperature. Then I would be ready with the second set of plants that would be coming into bloom in a few days.

Nothing would make the petals open again, and I found that they would not open a second time until the plants had gone through a full night period of darkness without having their sleep interrupted.

Later, when the reserve flowers came into bloom, I exposed all the individual frames of the petals in any one particular stage of development in quick succession. It was necessary to skip a number of frames and leave them unexposed until the flower reached various stages of development. Then I would go back and fill in the unexposed frames and thus be able to give the effect of the flower opening and closing whereas actually the petals only opened once and would fade away without ever closing. In effect it would be like showing a picture of a flower opening on the screen and then reversing the projector and running it backwards again and repeating this same procedure to look as though the flower opened and closed several times. Instead, the same effect was obtained by photographing the proper stages of development in the respective locations along the film to record a number of openings and reverse action closings.

The method of taking these pictures is possibly of some interest, but the preliminary failure of this project of trying to speed up the opening and closing of the petals of flowers is of some possible importance in itself.

I believe it shows that the response of the petals in opening and closing is not of a mechanical nature but is tied into the principles of photochemistry. Certain chemical changes apparently take place during the dark night period while the plant seems to sleep. The petals will not open in response to the energy of the light until these chemical changes have taken place. I suspect that the plant produces a chemical substance during the dark hours, but there is the possibility also that the plant could be disposing of certain wastes or byproduct chemical substances accumulated during the daytime. Either way there seems to be a close correlation with human sleep.

If the chemical produced by the plant during the night period could be isolated and produced synthetically, it would make possible an interesting experiment. Could this chemical substance be administered to the plant in such a way that the petals would respond to light repeatedly without the plant having its uninterrupted night sleep?

If so, then would the same principle work with animals and humans? If this proves to be the case, then it might add helpful knowledge regarding such drugs as tranquilizers. Carrying this even further, it might even be possible some day to make a pill that would be the equivalent of a good eight or twelve hours sleep. Sounds incredible, but no more so than a trip to the moon seemed only a few years ago.

Five

LIGHT AND THE ENDOCRINE SYSTEM

A happy accident stemming from my TV appearances now occurred and it came in the form of a letter from a viewer. He was a biology teacher in Chicago doing research with fish eggs and he wanted to experiment with time-lapse pictures. Delighted to help, I moved one of my time-lapse units right into his laboratory. Acting on a hunch, I suggested that we hang some of the various fluorescents used on the pumpkins over two or three of the fish tanks. Each fixture held two 40-watt fluorescent tubes placed about ten inches above the water and equipped with morning and evening timers.

Three different types of tubes were used—cool white, daylight white and pink. The aquariums were not located near any window, so the fish were being subjected entirely to fluorescent light. The first thing that happened was that the fish completely stopped laying eggs. Unhappily, because the teacher needed eggs for his work. On the other hand, it did indicate that light possibly did have some effect on the fish. After two weeks the light intensity was cut in half by removing one of the fluorescent tubes from each fixture. Still no egg production. Gradually we shortened the length of time the lights were left on. When the duration was down to eight hours a day, the fish began producing eggs.

Ordinarily, the sex of certain fish can be determined by the development of the secondary sex characteristics in much the same way brilliantly colored plumage is more noticeable on most male birds. We waited and watched the pinhead-sized fish grow larger day by day as the weeks went by. One day the biology teacher moved all the aquariums and lights from their location alongside his desk to another room. Several days later he told me that the young fish were all beginning to look suspiciously like females but it was still too soon to be certain. We waited approximately a month past the time the sex can normally be determined, and one evening he called me and said he had carefully checked each fish. He couldn't believe it, but all the ones under the pink light were definitely females. This was just the opposite from what might have been expected after the results with the pumpkin flowers. We waited another ten days, and then I told a few people about this first experiment with the fish. Although this was only one incident, and could not be given any scientific significance until repeated many more times, it was interesting that all fifty fish hatched from eggs of different parents appeared to be female.

The very next day after telling the story of the female fish, the phone rang again, and the biology teacher was quite excited—or possibly I should say upset—for now some of the fish he was certain were females were beginning to show faint signs of masculine coloration. We waited several weeks more, and the final results were that 80 per cent of the fish definitely turned out to be females and 20 per cent were questionable. They appeared to be males, but the development of the secondary male sex characteristics had been materially retarded.

Word of the preliminary results reached the newspapers, and articles appeared in several magazines. As a result, a letter came from a lady in New Jersey who wrote as follows:

My sixteen-year-old son, who has a very keen interest

49

in science, drew my attention to an article in a magazine regarding your experiments with light rays and their possible effect on sex determination.

I happen to be a chinchilla breeder and at the present time I am trying to establish a sizeable herd with sufficient breeding animals to enable me to start pelting within the next few years.

The loss of two excellent producing females in the last eight months, plus the fact that for the past three years my few breeding animals have yielded one female and nine males prompts me to write this letter to you.

I realize that perhaps you are not too familiar with chinchillas, but females are at a premium since one good male can serve several females thereby increasing the herd more rapidly plus cutting down on the costs of feeding, cages and space required. My interest in your experiments is more than passing, since I am in a position whereby I could benefit greatly if it were possible to produce more females than males.

Would you advise experimenting with my breeding stock, on my own, of course, as I certainly wouldn't expect you to do it? I am not interested in learning or obtaining your "trade secrets" and if you should advise in the affirmative, I would initiate the program on my stock only after I had consulted with my veterinarian since it would be too expensive a gamble otherwise.

Chinchillas have very dense fur and are very sensitive to extreme heat, thriving best in cool, dry temperatures of between 65-70 degrees. Even in this temperature, chinchilla books state that they can be overcome with heat if they happen to be in a cage where bright, warm sunlight strikes them through a window for several hours during the day....

Right now I imagine you have concluded that there is a selfish motive behind my interest and I suppose basically I would have to admit that there is, but the majority of small ranchers are faced with the same problem as I, more or less, and are being "held back" because they lack enough females to increase their herds rapidly.

It is very, very discouraging to wait anxiously for three and a half months for a litter to be born only to discover that the offspring are all males. And then you wait another three and a half months and find all males again. At this stage of the business, a rancher is too small to pelt these males off and doesn't have the females to mate them with so he reaches a stalemate which puts him behind several years insofar as realizing any profits is concerned. The excess males cannot sire offspring and increase the herd, but they still require cage space and food....

I wrote and asked for all the details and particulars regarding the lighting conditions of the location where she was keeping her animals. She advised me that she was keeping the chinchillas in cages in a basement play room. The room had one ceiling fixture with a regular 75 watt incandescent light and a small window at one end of the room. She also advised me that one baby female chinchilla was born of the animals in the cage at the end of the row nearest the window.

These conditions seemed to match closely the pumpkin situation except that she was using ordinary incandescent light whereas I had used fluorescent tubes to supplement the sunlight on the pumpkins. I purchased two 100-watt daylight incandescent bulbs and sent them to her, one to be used and one as a spare. This is the kind of bulb with the clear bluish glass that you can see through compared to the painted, frosted type that makes it impossible to see the filaments.

At last I received another letter from the lady in New Jersey:

The blue light bulbs arrived on November seventh, and I want to thank you for them.... The first litter just arrived on January third. I am not in the habit of handling new babies until they are a week old unless it is absolutely necessary, but yesterday I couldn't control myself any longer, and I still find it hard to believe that I found three female baby chinchillas....

Again this was only one isolated instance, but a very interesting one. Another matter of particular interest was that the blue lights did not arrive until somewhere between one-third and one-half way through the period of gestation. If the lights did have anything to do with the controlling of the sex of the baby chinchillas, it would indicate the controlling factor had to do primarily with the female parent. It would also indicate that the sex of the offspring could be influenced well along during the pregnancy. Several months later another letter came advising that the next litter from other parents was all females. This was doubly interesting.

Once, while reflecting on some of the unusual results that seemed to be associated with light, it suddenly occurred to me that the various growth responses that I had produced or controlled were from using different types of artificial light containing a peak of energy in specific narrow bands of wavelengths. The normal growth developments that I had prevented, such as the apple not ripening, were the result of filtering certain wavelengths from natural sunlight. This positive and negative way of acting certainly emphasized the importance of specific wavelengths of light.

Possibly a picture in one of my films might hold an answer or at least a partial explanation of all this. I found a microscopic time-lapse picture showing the streaming of the protoplasm within the cells of a living leaf of a plant. This activity goes on in connection with the process of photosynthesis in which the leaves respond to the energy of sunlight. They combine air and water with the minerals taken by the roots from the soil to create the food energy that supports all life on this earth. When the sun sets and it gets dark, this process of photosynthesis stops. It is a process of photochemistry through which chemical changes take place within the cells of the leaves of plants as they produce chlorophyll, carbohydrates and other chemical substances. Inasmuch as light is the source of energy that brings about these different chemical changes, it then seems reasonable to assume that by changing the characteristics of the light, the resulting chemical changes would likewise be altered.

This could explain the control of plant growth in response to the wavelength energy of light, but how could light affect the fish and the chinchillas? Here, the fact that the poultry industry knows that light received through the chicken's eye stimulates the pituitary gland and increases egg production might be a very important clue. The pituitary gland is the master balance wheel of the entire glandular system, not only in chickens but in other animals and humans as well. If this is so, and the entire glandular system can be affected—or glandular actions modified—by light received through the eye, the resulting consequences and possibilities of what this might mean are utterly fantastic. Possibly the basic principles of photochemistry in connection with the process of photosynthesis do carry over from plant life into animal life, but in a greatly improved way. If the basic chemistry of the human body responds to glandular actions controlled by the pituitary gland responding to light energy, then—as with plants—the characteristics of the light energy would be a very important factor. Different types of light and lighting conditions ranging from natural unfiltered sunlight to various kinds of artificial light, or natural sunlight filtered through different kinds of glass, or light reflected from different colored interior decorations in a room could affect the physical well-being of an individual.

It occurred to me that two films I had made might possibly offer further clues to the effects of light on the basic chemistry of both plants and animals. One was *The Story of Wheat,* made for the Santa Fe Railroad, covering the growing, harvesting, transportation and marketing of wheat. The other was a study of tomato virus for the Wright Brothers Greenhouses in Toledo, Ohio.

In searching for a proper wheat field, the County Agent called our attention to two wheat fields on opposite sides of the road, both belonging to the same farmer. One field was outstanding, the wheat waist high, with large, full, firm heads. The other, badly infested with virus, was ankle high. The virus-ridden field had been replanted regularly to wheat for years without any crop rotation or application of fertilizer. Obviously, there was a nutritional deficiency in the soil and the County Agent explained how this lowered resistance and made the crop more susceptible to disease. What impressed me most was the uniformity of the virus throughout the field. There wasn't the slightest trace of virus in the healthy field across the road.

The other half of this story pertains to the studies of tomato virus. The Wright Brothers have fourteen acres of tomatoes growing under glass in Toledo, Ohio. There are a great many hothouse tomatoes grown in northern Ohio. The tomato virus is one of the biggest problems growers have to contend with. It usually appears following long periods of cloudy weather and low sunlight intensity during the short winter days. It breaks out even under the most sterile and carefully guarded conditions. Nevertheless, it is generally agreed that the low light level also weakens the plants so they become more susceptible to attack from the virus.

During the course of making the film, I brought some virus-ridden tomatoes from the glass greenhouse into my plastic greenhouse. Ordinarily, such plants are rogued out and burned immediately before the virus can spread. The plants always seem to die anyway. With just a few days of sunlight in my greenhouse, and a light foliar feeding of the leaves, the tomato plants quickly came to life, started new healthy growth and began producing normal tomatoes.

In what I have learned about viruses, no consideration has been given to the possibility of a virus originating within the living cells of the plant itself. It seems to be generally accepted that the virus must be introduced from an outside source.

The metabolism, or life itself, that goes on within a living cell is the utilization of the nutritional factors present by the energy of light. The nutritional factors are like the coal or oil used for fuel to fire a boiler, and the light energy could be compared to the fire that burns it. Another comparison would be the gasoline used in an automobile engine and the spark that ignites it. If the draft in the boiler is not adjusted right, or the carburetor is giving too rich a mixture, there will be incomplete combustion. This can result in both the boiler and engine giving off not only obnoxious smoke and fumes but also partially consumed fuel. In a similar way, it seems quite possible that a chemical substance of a poisonous nature could result as a by-product from an incomplete or unbalanced metabolism within the cells of a leaf. This could result from either a nutritional factor as in the case of wheat virus or light deficiency as with the tomato virus. If so, then this chemical by-product would fit all the various descriptions of a virus. It would not be capable of reproducing itself, but if injected into the cells of other leaves, it might throw the metabolism of these cells off balance so that they would in turn produce more of the same chemical substance of a poisonous nature. It could be easily transmitted from one plant to another either by direct contact or some intermediary carrier. It could also be isolated and crystallized. It could fit all the various descriptions of a virus and still originate within the affected plant itself. This might also explain why too much plant food will kill a plant faster than not enough—simply too much of a good thing.

By now, a new theory was boiling within me and I determined to attack the virus problem through time lapse. To take pictures that would show what I wanted to study required building additional time-lapse equipment specially designed for taking microscopic pictures. This new unit was designed to take microscopic time-lapse pictures of the streaming of the protoplasm within the cell of a leaf as stimulated by direct unfiltered sunlight, as contrasted with various types of artificial light illumination.

The temperature of the subject being photographed could be controlled from 0° to 250° Fahrenheit. A secondary optical system was designed to superimpose an image of an electronic thermometer recording the temperature so that it would show simultaneously with the picture of the subject. In this way it would be possible to show precisely the effect of different sources of light and variations of temperature on the photochemistry that goes on in connection with the process of photosynthesis within the cells of a leaf. It would then be possible to study the effect on the germination of spores, mitosis of cells and other growth processes.

Six

I BREAK MY GLASSES

The increasing demand for time-lapse pictures necessitated building a mezzanine floor in the plastic greenhouse and installing additional cameras. More and more bits of interesting information were turning up at a much faster rate. All in all everything was beginning to run more smoothly except for two major problems. First, no one would pay any serious attention to the medical research possibilities of my time-lapse films. Second, advancing arthritis in my hip was making it increasingly difficult to carry a projector around for lectures or even to go up and down the basement stairs. Several doctors had recommended wearing a large metal brace and advised that a plastic hip joint would be necessary before very long. As a result, my wife and I were seriously considering moving to a house on one floor in order to avoid the stairs.

But the time-lapse studio created a real problem, for this would be extremely difficult and costly to move. Meanwhile two lecture trips took me to Florida during two successive winters. While there I spent as much time as possible on the beach to find out if basking in the sun would possibly help my arthritis. There were many stories about arthritis being affected by weather, but much as I would have liked a good excuse to spend more time in Florida during the winter, I could not honestly notice the slightest benefit. Sometimes it actually felt worse. On one trip I drove the family down and back. I enjoyed driving in the country, but my arthritis was always noticeably more aggravated at the end of the day regardless of how comfortable and relaxing the driver's seat might be. Nerves and fatigue, said the doctors. I should relax more. But how could I relax more than by sitting in the sun on the beach? Furthermore, while driving the car or sitting on the beach, I was always extra careful to wear my dark glasses to avoid any eye strain, since my eyes were very sensitive to the bright sunlight.

The only other times my arthritis definitely seemed to bother me more were immediately after my regular weekly TV program and following its filming the next day in the converted garage studio. Here maybe I could agree with the doctors about nerves and fatigue, but some extra aspirin would usually help considerably. Other than this, it was not possible to correlate my arthritic discomfort with anything else, including diet. Many well-wishing friends brought various remedies, tonics, and vitamin pills that had cured some distant relative. My arthritis must have been of a different variety, as none helped at all. Hot baths were relaxing but of no real value. Injections of various new glandular extracts would increase the discomfort for the first day or two and then give only four or five days' relief. Then the arthritis would be right back again. A cane helped a great deal by relieving some of the weight from my hip, but after using it for over two years, my elbow began to give trouble. I rode a bicycle around the yard back and forth between the house, tool shed and greenhouse. It was a girl's bicycle because it was easier to get on.

The problem of what to do continued to become more acute; then, one day I broke my glasses. While waiting for a new pair to be made, I wore my spares. The nose piece was a little tight and bothered me, so I took them off most of the time. The weather had been nice for several days, and there was some light work outside that I did as best I could with my cane in one hand. Suddenly I didn't seem to need the cane. My elbow was fine and my hip was not bothering me much even though I hadn't taken any extra amount of aspirin. It was hard to figure out why my arthritis should suddenly be so much better. My hip hadn't felt this well for three or four years. I began walking back and forth on the driveway. Fifteen minutes went by, and I must have walked a mile. I ran into the house and up the stairs two at a time to tell my wife. She had been watching me out the window and worrying. Had I lost my cane again? And why all the walking back and forth and around in circles without my glasses? It was shortly before Christmas, and—I told her—if she would hurry and finish her Christmas shopping, we could go to Florida for a week between TV programs. I wanted to sit in the sun again without any glasses. In three days we were on a plane headed south.

During that week the weather was very cold; in fact, an overcoat was necessary most of the time. Nevertheless, it was possible to be outdoors in the natural sunlight all day without any glasses. Perhaps this was a good thing because the light intensity away from the beach was not as great and made it much easier to do without dark glasses. At all times I was careful not to strain my eyes from too much light and never looked directly at the sun unless it was quite hazy or a little cloudy. I was also careful to guard against sunburn. Much of the time was spent sitting under a palm tree where I could read or look out into the open and still receive the benefits of natural sunlight in contrast to artificial light or sunlight filtered through glass. Fortunately, I was able to read without my glasses, needing them primarily for distant vision. My particular reason for not wearing dark glasses was that in addition to the glass itself filtering out virtually all the ultraviolet and certain other shorter wavelengths of sunlight energy, the characteristics of the light are further changed depending on the color of the glass. This acts as a filter restricting the transmission of all the other colors or wavelengths and transmits a peak of energy of the particular wavelength of whatever color the glass happens to be. While in the hotel, it was a great temptation to look out through a big picture window at the tropical vegetation and beautiful blue ocean. Conscientiously though, I avoided looking through the window glass and drove as little as possible to eliminate looking through an automobile windshield. I avoided bright artificial lights and did not watch television or go to the movies.

The effect on my arthritis was as beneficial as an injection of one of the glandular extracts right into the hip joint, but without the intervening day or two of increased discomfort. There was no doubt about it. My arthritis was definitely much better, and I was satisfied it was not imagination or wishful thinking. Furthermore, after several days of not wearing glasses at all, my eyes were no longer so extra sensitive to the bright sunlight even on the beach. Before the week was up, I played several rounds of golf on a short nine-hole course and went walking on the beach without my cane. I felt like a new person.

Theories may be interesting to think about and discuss with other people, but this was affecting my own arthritis, a much more personal and realistic matter. Maybe I was one of the lucky people you hear about who get better for no reason at all, but I felt strongly that there was a reason. I had taken my glasses off and let the full unfiltered natural sunlight energy into my eyes and had also made a point of being outdoors six hours or more each day whether it was sunny or cloudy. To me the results were convincing enough: that light received through the eyes must stimulate the pituitary or some other gland such as the pineal gland about which not too much is known.

The pineal gland is present in all craniate vertebrate animals. It is thought to be a remnant of an important sense organ utilized to a greater extent by more primitive animals. It is in most cases located at the base of the brain, but with some fish and reptiles—and especially certain lizards—it is raised near the upper surface of the head and has the structure of an eye with a more or less distinct retina and lens. It is then called the pineal eye. At any rate, something was stimulating the glands that lubricated my joints without artificially injecting any of the prepared glandular extracts.

Back home, I continued to stay outdoors every day without my glasses as much as possible from before sunrise until after sunset in spite of cold or cloudy weather. I used a small blue Christmas tree light as a night light in the bathroom just in case momentarily interrupting the dark night period of human sleep with bright artificial light might possibly have some detrimental or adverse effect as it definitely did with so many different plants. I moved my office from a room in the basement that had nothing but artificial light to a corner of my plastic greenhouse. When it was warm enough to be outside, I did as much office work as possible right out in the open. I also went swimming a great deal or otherwise wore a bathing suit as much as possible. For over a year I had spent almost two full days each week under the bright studio lights in order to repeat my weekly television program so that it could be recorded on color film. The total time under the intense studio lights was therefore cut from approximately sixteen hours a week to forty minutes at the most. This in itself made a tremendous difference, but even so, my arthritis still noticeably bothered me after each television program or driving a car for any considerable distance and looking through the glass windshield.

Theoretically, if this theory of light energy affecting the basic body chemistry is right, then it might go even much further as far as being responsible for various ailments and diseases, particularly of the old age or degenerative type, but all this needs further extensive study before any positive statements can be made. A friend whom I told of my experience undertook the same regimen and his hay fever vanished. Could not wearing glasses and being out in the sunlight possibly bring about a change in the body chemistry so the grains of pollen remained dormant?

One day I met a man who had previously taken a number of still photographs for me. He had meanwhile been on an assignment that required an intense amount of artificial lighting in large interior areas. He was an extreme diabetic and while on this job had a severe attack which resulted in the bursting of some blood vessels in the retina of both eyes. He became almost totally blind and could just distinguish the difference between day and night. He had been in this condition for approximately four years during which time he had numerous additional blood vessels burst in his eyes. He continued to work for the same company but in the photographic dark room where he was put in charge of processing film. Between batches of film he would occupy his time by reading Braille—in the dark.

The day I saw him again and learned of his blindness, I told of my experience with arthritis. Arrangements were made with his boss for a table outside where he could read his Braille while waiting for films to be processed. He made an effort to be outdoors as much as possible while at home. Approximately six months later, he had not had a single blood vessel burst, could distinguish different colors, and see enough to follow the vague outline of the sidewalk ahead as he walked to work. Another single isolated case, but a very interesting one to follow. Incidentally, he always wore thick, strong glasses before going blind.

Some doctors have said cancer is caused by a virus or at least is in some way associated with it. If this is so, then the possibility of influencing body chemistry by the characteristics of the light energy received through the eye might conceivably be an important factor in the metabolism of the individual cells of the tissues of the body. The same principles of nutritional factors, light energy, and a balanced metabolism would follow the same line of reasoning as with both the wheat and tomato viruses.

Just exactly how this energy could be transmitted was hard to visualize. Nevertheless, I had photographed plants that could transmit energy or impulses quite rapidly, and certainly these plants have no nervous system similar to that of animals or humans. Both the venus fly trap and sundew plants are good examples, but even better possibly would be the *Mimosa pudica*, or sensitive-plant. An interesting characteristic of this plant is that it folds its leaves tightly together when it gets dark and seems actually to go to sleep at night. It opens its leaves again during the daytime. If you touch the leaves with your finger or strike them with any object, they quickly fold up in about one second. If the plant is left undisturbed, the leaves will slowly open again in approximately five or ten minutes. If the tip end of a leaf is singed with the flame of a match, the shock is greater, and the reaction can be seen as it travels throughout the entire plant. The singed leaf first folds up quickly, then the branch collapses. The shock wave travels in one or two seconds through the main stem to the other branches which collapse. Then the shock continues to travel through these other branches to the leaves that finally fold up. Again, if the plant is not disturbed, the leaves will slowly open in approximately ten minutes time.

A further interesting phenomenon is that the entire plant can be anesthetized with ordinary ether so that it will not react even to the more severe shock of singeing a leaf with the flame of a match. This may be done by placing some cotton saturated with ether near the plant and covering it over with an air-tight cover. When the cover is removed and the plant has been in fresh air again for ten or fifteen minutes, it will react in its normal way. Another interesting observation regarding the sensitive-plant is that even though it is kept in total darkness in the basement under a concrete ceiling, as well as the usual concrete walls, the leaves continue to open and close according to the outdoor daylight or night periods. Whether or not this reaction is controlled by cosmic rays or other radiant energy forces capable of penetrating concrete is something of a mystery.

The fact remains that these plants are capable of transmitting this energy or shock impulse in a way that is not fully understood. Therefore, it seems reasonable that light energy or the effects of it could be similarly transmitted through animal tissue and become an important factor in the metabolic function of the individual cells.

I showed my pictures and stressed the effects of light and its important possibilities to a number of medical groups, universities and the research personnel of seven large pharmaceutical companies across the country. Same reaction every time. Excellent pictures, very interesting, and somebody would be getting in touch with me. But nothing ever happened. One company wanted to test out the theory of spores in connection with the common cold but was unable to find anyone with a cold at the right time after searching for six months in the New York City area. Another company was interested in helping with some of my projects until

I suggested they also help by sharing some of the expenses. This abruptly changed its attitude. I showed my pictures and told my story at the headquarters of the United States Department of Agriculture, Public Health Institute, and Surgeon General's Office of the Army—all at my own expense—but could stir up no action. Finally the United States Information Agency became interested and translated the most recent and complete magazine article about my work into Russian and sent it to Moscow. I tried to interest several of the large foundations, but with no results. Two universities made an appeal for a research grant based on time-lapse photography that I would do jointly with them, but were turned down cold.

The research departments of several large corporations showed some interest in the possibilities of time-lapse photography. However, their interest was only in its application to particular problems on which they were already working. The heads of the research departments of two other large companies confidentially expressed some interest but frankly said any official recognition or participation in such an outlandish idea would subject them to the risk of possible ridicule by other scientists. Invariably, they would all check the available literature on the subject and report there was nothing to support my observations. The information in the literature dealt primarily with color therapy and the psychological effect of different colors on more or less emotionally unstable people. It was no help at all and only tended to classify me further in the category of crackpot.

Progress was slow and discouraging. At times the whole idea seemed utterly ridiculous even to me, and often quite hopeless. Many times I pinned my hopes on a particular showing of my films for some official recognition and acceptance of the importance of light energy and other interesting phenomena revealed through my time-lapse pictures. There was always the same polite but negative response. Several senior educators and top doctors suggested quite frankly that I should forget about any medical application or reference in connection with light, particularly concerning cancer, before I brought too much ridicule and disgrace not only upon myself but my family as well.

Then, things began to happen that were tremendously helpful. The Chicago Technical Societies Council honored me with one of their annual Merit Awards, "for outstanding technical achievements, service to science, fellow scientists and the community." Soon afterwards Loyola University in Chicago conferred upon me an honorary degree of Doctor of Science. Next I was asked to become a member of the faculty in the Biology Department, and very soon after this I was also made a member of the faculty of Michigan State University in the Department of Horticulture. Meanwhile members of the faculty of other universities including Harvard, Illinois, Iowa, Northwestern, Purdue and Wisconsin cooperated wholeheartedly with me in an advisory capacity with the production of various technical films for a number of nationally known large corporations. The Chicago Horticultural Society awarded me the Charles L. Hutchinson Medal for my "time-lapse work in horticulture and contributions to the scientific knowledge of plant growth." All these associations helped tremendously in lending scientific dignity to the theories I was postulating.

The idea of using time-lapse photography for more than simply entertainment and advertising films was taking hold. Additional experiments were started by others at both Loyola and Michigan State Universities. One of the most encouraging and gratifying experiences came when the Lamp Development Department of the General Electric Company retained me as a consultant to study and advise on the effects of radiant energy on plants and animals. The Quaker Oats Company placed their research farm near Barrington, Illinois, at our disposal. Experiments were started in subjecting chickens and various domestic animals to different types of lighting conditions. The Quaker Oats Farm was otherwise used primarily for testing various animal feed formulas. This was a tremendous help, since it was already established as a well organized and smoothly operating experimental farm.

More recently I have built two time-lapse camera units for use in the Cancer Research Program at Chas. Pfizer and Co., Inc. They have retained me to consult with them on time-lapse problems and the manner in which they relate to my own work. Other important companies have also indicated an interest in this subject of the importance of the full spectrum of sunlight energy, and a number of exciting experiments are either already started or in the definite planning stage. These experiments may take several years to complete and will undoubtedly lead to other experiments requiring additional years. This raises the question of whether or not the best policy would be to keep all information strictly confidential and release nothing until scientifically proven beyond any doubt. On the other hand, it is my firm belief that by making as much information as possible available—with caution—that others may possess information that might supply the missing pieces to the overall puzzle.

Meanwhile more bits of interesting information keep turning up from the most unexpected sources. On one of my regular television programs, I was privileged to have Warden Joseph E. Ragen of the Illinois State Penitentiary as my guest. His work in the rehabilitation of men at Stateville Penitentiary and the importance of horticultural therapy had been written up in one of the Chicago newspapers. This sounded most interesting, so I contacted Warden Ragen and was invited to see the prison gardens and work being done along these lines inside the prison walls. The extent of the gardens and their beauty was simply amazing. The fact that the men did all the work and raised the plants themselves was certainly commendable, but their obvious enthusiasm and feeling of personal pride in their work impressed me most. Warden Ragen showed me many letters received from men after their release as well as letters received from men still at Stateville that might be summed up by the remarks of one man who came there as one of the toughest criminals and psychological problems ever to be dealt with. Warden Ragen told me he had stopped one day and asked this man how he was getting along. The man straightened up, pointed to the flower bed he had just finished cultivating and said:

> Warden, this is the first decent thing I've accomplished. I've been a thief and criminal all my life. All my gains were ill-gotten, and I find now I can do something that will be worthwhile, not only for myself, but for people as a whole. I know flowers are not only pretty, but they're profitable as well. I'm sure that when my sentence is served, you'll never hear from me again so far as crime is concerned. I'm going to ask you to help me find employment in a greenhouse or as a gardener.

From later correspondence I had with Warden Ragen, I again quote from one of his letters as follows:

> I should like to say one thing, and one which can possibly be considered repetitious on my part. I am positive that schools for delinquents, reformatories and prisons are not the proper place to make good citizens. I do not

think that children are instinctively born criminals. I believe they are led into lives of crime in many instances, by delinquent parents, improper home situations, lack of love and care to which they are entitled as children and lack of religious, academic and vocational training. Certainly, if our prison populations are to be reduced, we must do more about the "cause" which produces the delinquent child of today. He must be guided through his formative years on the road to good citizenship rather than be permitted to drift to a life of crime and disgrace, and further, become a very expensive liability to taxpayers and society in general.

Certainly working with flowers and plants in the garden close to nature is a very good psychological influence. Possibly being out in the sunlight as contrasted to the solitary confinement in a dark or artificially lighted cell is even more basically a good thing. If it is the natural sunlight received through the eyes that is beneficial in helping to rehabilitate such men, then it might also be an important factor in juvenile delinquency. In my mind it raises the question about the ultimate effect on human health and normal growth development resulting from excessive exposure to other than natural sunlight as the result of increased and extensive use of large picture windows, glass buildings, and modern brighter artificial lighting. The fad of wearing dark glasses is sweeping the country. The matter of driving or being driven in an automobile to school or work is becoming more important all the time. Many outdoor sports are now attended at night under lights or watched over TV. The importance of education is being stressed more and more, and students are working harder and longer under midnight electricity to meet stiffer requirements and increased competition. New mental institutions, hospitals and especially maternity wards, where newly born infants get their first glimpse of light, have larger windows that are no longer made to open—and more and brighter artificial lights.

For several years I had been increasingly bothered with common head colds and a sore throat. Several people who regularly watched my TV program either sent me or recommended various cough remedies that seemed to have little effect. This troublesome condition also disappeared as I continued to practice my theories of being outdoors in the natural sunlight. For some time I more or less joked with various friends including some in the medical profession about feeling so much better, and all agreed wholeheartedly that it was a wonderful thing regardless of whether it was due purely to my imagination or not.

It might be well-noted here that after six months of not wearing glasses, except for what little driving of the car was absolutely essential, and for focusing my projector when showing pictures, I began to notice that wearing my glasses even for these short periods seemed to strain my eyes more and more. Accordingly, an appointment with my oculist for a regular check-up seemed advisable. This time it was necessary to go back for a second examination which my doctor explained was customary in order to double check any such drastic change as was the case with the condition of my eyes. The principal difference in my new prescription was that the rather strong prisms previously needed to correct a muscular weakness were no longer needed. With this encouragement, I decided also to have my hip X-rayed again.

It was most gratifying to have my doctor advise that the X-ray pictures showed a definite strengthening and improvement in the area of my hip joint that had been causing so much trouble. A physical examination revealed the complete disappearance of a 30 per cent restriction of the movement or rotation of the hip joint which my doctor commented on as being wonderful but quite surprising and most unusual. For six months I had been imagining I felt better, and it was a great relief to have these X-ray pictures and examination confirm my imagination.

Seven

AN EXPERIMENT WITH PHOTOTHERAPY ON HUMAN CANCER PATIENTS

The initial hope for the opportunity to work on some carefully planned and scientifically controlled light research experiments with the experienced personnel and sophisticated facilities of several of the largest corporations in the country rapidly faded into disillusionment. The positive results obtained with the various pilot experiments using fish, chickens and chinchillas produced, in each instance, a reaction of doubt and a suspicion that something not much short of witchcraft was to be suspected. To suggest that light entering the eyes could have any biological function other than producing vision was like seriously talking—twenty years ago—of man's foot prints someday being on the moon. A reexamination of each company's research policies, procedures and financial budgets indicated that no provision existed for the study of the effect of light on animals. Furthermore, nothing could be found in the literature to support the hypothesis. This was not the first such disappointment, nor was it to be the last.

However, there was also the bright side of things. Following my last visit as a consultant to the Chas. Pfizer Cancer Research Laboratory in New Jersey, I was offered a ride back to New York City in one of the company's chauffeur driven cars. A physician, Dr. Jane C. Wright, in charge of cancer research at Bellevue Medical Center in New York City was in the same car. We started talking about cancer research and she expressed interest in the time-lapse pictures and the suggestion that there might be a relationship between light energy and viruses and the increasing interest in the cancer virus theory. To my delight, she agreed to ask fifteen cancer patients to spend as much time as possible in natural sunlight without their glasses, and especially their sunglasses. They were also instructed to avoid artificial light sources as much as possible, including television. This experiment was conducted during the summer months of 1959.

At the end of the summer, Dr. Wright advised that while it was difficult to make a positive evaluation, it was the consensus of all those assisting in the program that fourteen of the fifteen patients had shown no further advancement in tumor development and several showed possible improvement. The fifteenth patient had not fully understood the instructions and although she did stop wearing sunglasses, had nevertheless continued to wear ordinary glasses which would of course block most, if not all, of the ultraviolet in natural sunlight from entering the eyes.

As cold weather was approaching, it would not be possible for these people to remain outdoors much of the time in the New York City area. So Dr. Wright made arrangements for me to show the time-lapse pictures and explain the story again to the general research staff of the M.D. Anderson Hospital and Tumor Clinic in Houston, Texas, on January 27, 1960. It was hoped that in view of the results obtained with the pilot experiment in New York, a greater number of patients might be kept outdoors under natural daylight conditions without their glasses the year around in a southern climate with a milder temperature.

However, as I presented my story I became aware that the atmosphere was becoming progressively colder and, in fact, the general response, even before I had completed the story, was stone cold. Dr. Wright, in New

York City, started to make plans to repeat the experiment there the following summer, but shortly before the project was scheduled to be started, I received a letter stating that circumstances made it necessary to call it off. In fact, criticism of the project had been so great that it seemed advisable not to make any further mention of the previous year's experiment at all. The main objections were that no patients were actually used as controls and that any such experimental procedures should be first proven with animals.

Although I was careful not to mention the human cancer experiment in my various letters, I did give all the information to another physician in the Chicago area who had been a close personal friend for many years. He was Dr. Samuel Lee Gabby, Senior Staff Member, Sherman Hospital, Elgin, Illinois. He agreed to set up some experiments using the C_3H strain of mice, which is highly susceptible to spontaneous tumor development, and subjecting them to different lighting environments. I assisted him in setting up the equipment and maintained a close working arrangement throughout the experiment. Thirty pairs of test mice were kept in a room lighted only by daylight white fluorescent tubes. Thirty pairs of test mice were kept in another room lighted only with pink flourescent tubes, and, as a control, eight pairs were kept in a room where they received daylight filtered through ordinary window glass. The control mice in the daylight cages developed cancer some two months later than the test mice which were kept in the room with the daylight white fluorescent tubes, and three months later than those under the pink fluorescent lights. The different lighting conditions also noticeably affected the litters born during the experiment. The litters of mice under pink fluorescent light consisted of only one or two offspring instead of the normal six to fifteen under the daylight white fluorescent, or under outside light coming in through the ordinary window glass. The full text of this report was submitted to the *Illinois Medical Journal,* but unfortunately was not published. It was unofficially presented to several members of the Illinois branch of the American Cancer Society reviewing committee, but did nothing more than stir up further criticism of my good friend, Dr. Gabby, who had collaborated with me.

Eight

CHLOROPLASTS AND LIGHT FILTERS

The problems encountered in taking time-lapse pictures continued to open up intriguing possibilities for further research on the effect of light on both plants and animals. I had already made personal contributions to support this research and several corporations had contributed a few modest grants, but now there was need for more substantial funds to support a rapidly growing research program. I decided to incorporate the Time-Lapse Research Foundation on a nonprofit basis to carry on the research. Several prominent physicians and dentists agreed to serve on the Board of Directors. A grant application was carefully prepared with the assistance of several of the doctors who had had experience in making similar applications. We cited the limited work that had been reported in the literature regarding the effect of light on the retinal-hypothalamic-endocrine system in animals. We pointed out that light was an important part of man's total environment and that it could affect his overall general health and well-being. Perhaps it was a mistake to suggest again that the colors, or balance of wavelength energy of light entering the eyes might possibly influence the development or growth of cancer, even though no mention was made of the pilot experiment with human cancer patients. Our first grant application was dated February 26, 1962, and was forwarded to the National Institutes of Health.

The application was not approved and the reasons given were: The reviewers believed that the proposal did not indicate familiarity of the applicant with existing research in the field, or with scientific methods in general. They recognized that I was experienced in time-lapse micro-photography; that the films I listed in support of my request are of a very popular kind, but it did not seem likely that anything of scientific merit would emerge from such a program.

It became increasingly clear that somehow the story about the pumpkins, and the fish, chickens and the chinchillas would have to be published if anyone was going to take the matter seriously.

The first real break came in 1961, when I was invited to show time-lapse pictures and tell of working with Walt Disney on his nature film series as the main entertainment feature following a banquet of the New York Academy of Sciences. These unusual pictures were intended as relaxation, something quite apart from the serious two-day scientific program. Two members of the program committee told me this was the only way they could arrange for the people attending the scientific programs to see the time-lapse films. They gave me the go-ahead to present the full story of the effects of light on both plants and animals and agreed to watch for the reaction of some of the key scientific people attending the banquet.

I showed the time-lapse sequences of the pumpkin and the morning glories, the tomatoes and apples, and, of course, the dancing flowers. I also told of the experiments with the fish and chickens and the lady with her chinchillas. The plan worked, and I was invited back the following year to show the time-lapse pictures in connection with a paper I was to present to the scientific session of an international symposium on "Photo-Neuro-Endocrine Effects in Circadian Systems with Particular Reference to the Eye." This symposium was held at the Academy in June, 1963, and my paper was included and published as part of the *Annals of the New York Academy of Sciences,* Volume 117, Article 1,

September 10, 1964. Enlargements of the individual frames of the pumpkin sequence were used as illustrations. I was also able to include pictures of some additional work I had done in studying the response of individual chloroplasts within the cells of Elodea grass to different colors, or wavelengths, of light energy.

These studies showed particularly the importance of the near, or longwave ultraviolet that penetrates the atmosphere. The pictures showed that when the Elodea grass was exposed to the full spectrum of all the wavelengths of natural sunlight, all the chloroplasts would stream in an orderly fashion around and around from one end of the cell to the other. However, if the sunlight was filtered through ordinary window glass that blocked most of the ultraviolet, or if an ordinary incandescent microscope light, which is lacking in the ultraviolet part of the spectrum was used, some of the chloroplasts would drop out of the streaming pattern and remain immobile near the center or off in one corner of the cell of the leaf.

When a red filter was placed in the light source of the microscope, further restricting the wavelengths, more of the chloroplasts would drop out of the streaming pattern, and other chloroplasts would make a shortcut from one end, across the center of the cell, without going all the way to the other end as they would do when they were receiving all the wavelengths of natural sunlight. A green filter would extend the shortcut pattern from across the center of the cell a little further toward the far end and a few more chloroplasts would get back into the streaming motion. When a blue filter was placed in the microscope light, then still more chloroplasts would resume their streaming motion and those that were streaming would go almost all the way to the far end of the cell, shortcutting only a little across one corner.

When the color filters were removed and a low intensity source of near ultraviolet light was added to that of the regular microscope incandescent lamp in order to come as close to the full spectrum of natural sunlight as possible, nearly all the chloroplasts would go right back to their full, normal streaming pattern. But at the end of the day all of the chloroplasts would gradually slow down and remain virtually motionless, regardless of how much the intensity of the light source might be increased. They would not resume their normal streaming pattern until they had had their dark "sleep" period.

With reference to the various shortcut patterns caused by the different color filters, it is interesting to note that the red, or longest wavelengths, caused the greatest shortcut or variation from the normal pattern. Green wavelengths are near the center of the visible spectrum and a little shorter than those that we see as red and the shortcut pattern would go a little further toward the far end of the cell. Blue wavelengths made the chloroplasts go almost all the way around the far end, and adding the still shorter wavelengths of black light ultraviolet caused them to respond most closely to the way they do under unfiltered natural sunlight.

Showing those time-lapse films and speaking at that meeting at the New York Academy of Sciences did seem to break the ice. Even before the proceedings were actually published I was invited to show my films at the Fourth International Photobiology Congress at Oxford, England, during the summer of 1964, and also at the National Technical Conference of the Illuminating Engineering Society on August 31, 1964, at Miami Beach. My paper, presented at the Oxford Congress, was entitled *Some Observations on the Effect of Light on the Pigment Epithelial Cells of the Retina of a Rabbit's Eye,* and I was particularly pleased that this paper was one of four out of approximately 500 to receive an honorable mention during the final closing meeting of the Congress, and was also included in the official proceedings, *Recent Progress in Photobiology,* edited by E. J. Bowen, and published by Blackwell Scientific Publications, Oxford (1965).

The title of my paper presented at the I.E.S. meeting was *Effects of Wavelengths of Light on Physiological Functions of Plants and Animals,* and this too was published, in the *Journal of the Illuminating Engineering Society,* as part of the "Color Symposium" in the issue of April, 1965.

Both of these papers included the results of another interesting time-lapse project which came about as the result of visiting a retired uncle and aunt in Sarasota, Florida. My aunt belonged to a garden club that was in need of a program for their Men's Night meeting. She requested that I bring along one of my time-lapse films. I did, and it happened that a well-known ophthalmologist, Dr. Thomas G. Dickinson, was in the audience. After the meeting, he expressed great interest in time-lapse photography. In talking with him I found that he was particularly interested in the microscopic time-lapse pictures showing the Elodea grass cells.

Dr. Dickinson explained that there is a layer of cells in the retina of the eye known as the pigment epithelial cells that are thought to have no visibility function, and that while the purpose or function of these cells is not fully understood, it is known that they do show abnormal responses to the excessive use of certain widely used tranquilizing drugs. He asked whether I would be able to do a drug toxicity test utilizing microscopic time-lapse photography. It would attempt to show the effect of adding various dosages of these tranquilizing drugs to the growth media ordinarily used for growing cells in tissue culture chambers or slides that could be photographed through a phase-contrast microscope.

This was exactly the opportunity I had been hoping for—to utilize microscopic time-lapse photography in a project scientifically planned and under the direction and control of experienced research personnel. At this time I also had a new full-time assistant, Anthony Marchese, who had majored in biology and was interested in medicine. His hobby was photography, and this combination of interests proved invaluable.

Dr. Dickinson put me in touch with the research people at the Wills Eye Hospital in Philadelphia. Irving H. Leopold, M.D., Medical and Research Director of that hospital, came out to help start the project in our new laboratory at Lake Bluff, Illinois, where my wife and I had recently moved. It was not too difficult to take along the time-lapse cameras and microscopes, but to move the greenhouse to Lake Bluff was a bit more complicated. The head tissue culture technician from Wills, Rosemarie Nagy, also came out to Lake Bluff to work with Tony and show him how to prepare the tissue culture slides. Soon the research was going so smoothly, and the results were so satisfactory, that Tony Marchese was able to take complete charge of the time-lapse microscopic work. This project would be the next one to utilize the latest improved microscopic time-lapse unit containing two of the latest model phase-contrast microscopes.

Phase-contrast microscopes come equipped with a complete set of different colored filters for use in the built-in microscope light source. The mechanics of the optics of the phase-contrast microscope are such that sharper pictures with greater contrast may be obtained by using a monochromatic light source, which simply means a light source of one particular color. That is because there is a variation in the speed at which the wavelengths of different colors travel, and the simplest way to obtain a monochromatic light source is to put

81

different colored filters in front of the light, so that only the wavelengths of one color will be transmitted.

Green is the color most commonly used, but red and blue and other colors may also be used, depending on the nature and color of the subject to be photographed. The great advantage of the phase-contrast microscope is that the outline and details of the cell structure of the subjects being photographed can be clearly seen without having to stain the cells, as was necessary with the old type microscope. Staining the cells kills them, so that seeing the greater detail and being able to keep the cells living is of particular advantage in making time-lapse pictures over an extended period of time.

For some unknown reason the color filters were lost, or at least not received with the first phase-contrast microscope, so that the cancer cell pictures, made during the first such project for Northwestern University Medical School, were good, but not as sharp and clear as they might have been had I then known of the advantages of using a monochromatic light source, or color filter.

We quickly noticed that there were far greater abnormal growth responses in the pigment epithelial cells depending on the color filter used in the light source of the phase-contrast microscope than to the different tranquilizing drugs that were added to the growth media. This response was completely unexpected.

Exposure to blue light, or the shorter wavelengths, would cause abnormal pseudopodial activity in the pigment epithelial cells, while red light, or the longer wavelengths, would cause the cell walls to rupture and allow the cytoplasm to run out. The process of mitosis, or cell division, would not occur when the cells had been exposed to either blue or red light for approximately three hours or more, but only under a white light containing a more complete light spectrum. Fresh media is also important for mitosis, but adding fresh media to the slide chambers at constant incubator temperature would not encourage mitosis.

When the feeding of the cells with fresh media was done at normal room temperature and the tissue culture slides then replaced in the incubator, greatly accelerated mitosis would take place in approximately 16 hours. Toward the end of the normal daytime period, the activity of the pigment granules would noticeably slow down. Similar to the action of the chloroplasts in the plant cells, the pigment granules in the epithelial cells of the retina also required a dark period uninterrupted by light before resuming their normal response to light energy. This was another interesting similarity of responses in both plant and animal cells to the periodicity of light.

When I first started taking microscopic time-lapse pictures, I set the automatic timer to turn the microscope light on momentarily while each exposure was made. Individual exposures were then taken at regular intervals of 30 to 60 seconds on a continuous 24-hour basis. However, after the experience with both the chloroplasts in the cells of Elodea grass and the pigment granules in the epithelial cells of the retina of a rabbit's eye responding to a rest period without light, I found in many instances that there would be noticeably different results depending on whether the microscopic time-lapse pictures were made on a continuous 24-hour basis, with the microscope light turned on momentarily while each exposure was made, or whether the light was left on continuously during the daytime and turned off at night so that the cells being photographed would receive their normal dark rest period.

However, after a 12-hour continuous exposure to ordinary incandescent light each day for one week, an estimated 90 per cent of the pigment granules became sluggish in their action and remained virtually motion-less at one end of the cell. By adding the same very low intensity of black light or long wave ultraviolet to the ordinary incandescent light source that had made the chloroplasts get back into their full streaming pattern, all the pigment granules would become active again and move in their normal pattern within the cell. This would indicate that there are similar responses in both plant and animal cells to different wavelengths of light. No method of measuring the intensity of the ultraviolet was available, but optimum level had been determined by trial and error in the experiment previously men-tioned with the chloroplasts of Elodea grass. Increasing the intensity of the ultraviolet light, and especially adding additional short wavelength ultraviolet resulted in abnormal action of the pigment granules and death of the pigment epithelial cells within about 30 minutes.

As a result of these experiments, it is suggested that the responses of chloroplasts and pigment granules may be "tuned" to the natural light spectrum of sun-light, under which all life on earth has evolved. With respect to the ultraviolet range of wavelengths, the matter of intensity seems to be particularly critical. The normal intensity of the near ultraviolet, that is, the ultraviolet wavelengths longer than 2900A or 290 millimicrons at which point the earth's atmosphere filters out the shorter wavelengths, or far ultraviolet, may be an essential part of the natural sunlight spect-rum. The same intensity of far shortwave ultraviolet (shorter than 290 millimicrons), only a trace of which penetrates the atmosphere, is unquestionably extremely harmful.

It would therefore seem that the chemistry of plants may be affected by the various responses of the chloro-plasts to both the periodicity of light and darkness and to the intensity and distribution of wavelengths influencing the process of photosynthesis. It is further suggested that similar responses of the pigment granules in the pigment epithelial cells of the retina might be involved in the photoreceptor mechanism, referred to by some scientists without identification, that stimulates the retinal-hypothalamic-endocrine system (sometimes referred to as the oculo-endocrine system) in animals and thus influences the hormonal balance or body chemistry. And so it would appear that the basic principles of photosynthesis in plants, where light energy is recognized as a principal growth regulating factor, might be equally as important a growth regulating factor in animal life through control of chemical or hormonal activity.

In a paper published in the May, 1932, issue of the *Journal of Comparative Neurology,* Dr. Wendell J. S. Krieg, Professor of Anatomy at Northwestern University Medical School in Chicago, describes the retinal hypothalamic pathway of the albino rat and gives an excellent review of the literature from 1872 when Meynert first described the basal optic ganglia, or *supraoptic nuclei,* in man. Anatomists have long known of the existence of such glands as the pituitary and pineal, as well as many interconnecting neural pathways of the autonomic system. However, it has been only recently that endocrinologists and other interrelated disciplines have begun to explore what their purpose is and how they function.

In placing a filter of any particular color in a white light source, only the wavelengths of light representing that particular color are permitted to pass through the filter. On first thought it might seem that the resulting abnormal growth responses might be caused by the wavelengths of the color involved. However, these wavelengths that do pass through the filter are a part of the total spectrum of the original source of white

85

light, and the filter cannot add any additional energy to the spectrum of the original light source. It would therefore appear that any altered growth responses must be due to the absence of the wavelengths blocked by the filter, and that the lack of these wavelengths causes a bio-chemical or a hormonal deficiency in both plant and animal cells. This might be referred to as a condition of mal-illumination, similar to that of malnutrition.

Microscopic time-lapse pictures of other animal cells in tissue culture also showed similar variations in growth patterns when different colored filters were placed in the light source of the phase-contrast microscope. It was of interest to note how a red filter consistently caused the cell walls to weaken and ultimately rupture. This response was particularly noticeable when heart cells from a chick embryo were subjected to red light. This again raises the question of whether there may be any connection between coronary disorders and the high red content within the spectrum of ordinary incandescent light bulbs.

On two separate occasions, following the showing of these pictures, two prominent virologists commented that some of the abnormal biological effects produced by placing a blue filter in the microscope light source closely resembled the effects of cells being attacked by viruses. To me, this further indicates the possible relationship between the abnormal chemistry associated with viruses responding through the process of photosynthesis in plants and the retinal hypothalamic endocrine system in animals, to an incomplete, or unbalanced, light source. It was about this time that some medical scientists were suggesting that cancer might be caused by a virus.

When the pigment epithelial time-lapse pictures showing the effects of both the drug toxicity study and different colors or wavelengths of light were completed, I was invited to show them at one of the regular research seminars at the Wills Eye Hospital. I also showed the pictures of the pumpkins, apples, morning glories and tomatoes, and presented the complete story, including the experiments with the fish, chickens, and chinchillas. Now some of the preliminary results of the effects of light on laboratory animals such as mice, rats and rabbits could also be added. Dr. Irving Leopold was not only director of the Wills Eye Hospital but also editor of *Survey of Ophthalmology,* one of the recognized ophthalmological journals. He asked me to write a paper on the subject as I had presented it at the seminar. He thought the full story should be published in the journal. I did, and shortly thereafter I received the following letter:

July 7, 1961

Dear Mr. Ott:

I have had several members of the Editorial Board read over your material in the hope that they would accept it for the *Survey* but have had no luck so far.

I am returning it all to you and suggest that this be sent to the A.M.A. and ask them to suggest a journal which might use this information so it could be called to the attention of the doctors.

I am sorry that I was unable to bring this about as it seemed like such a good idea to me. The following is an excerpt from the comments of one of the reviewers:

"I cannot see that this subject matter belongs in *Survey* at all. The first 9-1/2 pages are pure plant physiology. The remainder has only the remotest of connections with ophthalmology. In some vertebrates having photoperiodism of their reproductive cycles, the retina may be the receptor in a quasi-reflex arc terminating in the pituitary or gonad; but the connection would be via one of the vague 'accessory optic tracts' and would have nothing to do with the *visual* system. For some vertebrates, it has long been known that photic control of reproduction is through direct stimulation, through the side of the head, of the photoreceptive ependymal lining of the third ventricle. A case in point is that of the duck, extensively researched for years by Benoit—to whose work Ott refers without any mention that the duck's *eye* has nothing to do with the timing of spermatogenesis, etc."

Best regards.

Cordially,
Irving H. Leopold, M.D.

After reading the last sentence, stating that the duck's eye had nothing to do with the timing of spermatogenesis, etc., as researched by Benoit, I reread the classic paper published by Jacques Benoit and Ivan Assenmacher (College de France, Paris) *The Control by Visible Radiations of the Gonadotropic Activity of the Duck Hypophysis*, as published in *Recent Progress in Hormone Research*, Volume 15:143-164, Academic Press, New York, and quote from it as follows:

As was established by Rowan on the Junco hyemalis, by Bissonnette on the sparrow, and by one of us on the duck, visible radiations induce an important gonado-stimulation in immature birds, or in birds in periods of sexual activity....

Through which paths does light stimulate the hypophysis? . . . They are effective on the eye, as was shown by experiments involving local application of light.... As for retinal involvement in gonadostimulation, it seems to function in a different way than for vision.... Moreover, a light stimulus is still very active, on the gonads after resection of both eyes or after bilateral section of the optic nerves. In the latter case, it could be established by both photographic and photo-electric methods that visible long-wave radiations (orange to red) penetrate deep enough into various tissues to reach and stimulate the hypothalamus. Thus we see that visible radiations can stimulate the gonadotropic function of the anterior lobe, both through an oculohypothalamic path and by a direct action on the hypothalamus through the orbital tissues.

My interpretation of Benoit's paper is that he distinctly states that the eye of the duck is the primary photoreceptor of light that stimulates gonadotropic function, but that after resection of both eyes or after bilateral section of the optic nerve, gonadotropic function can also be stimulated by direct action on the hypothalamus through the orbital tissues.

In similar context, 1970 Nobel Prize winner, Dr. Julius Axelrod, worked with Drs. Wurtman and Fischer on studies they reported in *Science*, Volume 143, pages 1328-1329, March 20, 1964, *Melatonin Synthesis in the Pineal Gland: Effect of Light Mediated by the Sympathetic Nervous System.* They state:

> Removal of both eyes resulted in a complete loss of the capacity of the pineal gland to respond to altered illumination with the accompanying changes in weight or HIOMT activity. This indicates that the action of light upon the rat pineal gland is not direct, but mediated by retinal receptors.

Dr. Thorne Shipley states in his article, *Rod-Cone Duplexity and the Autonomic Action of Light (Vision Research*, Volume 4, pages 155-177, May 1964):

> Thus not only is light itself of autonomic importance, but, confirming Benoit and Assenmacher *(1955)*, its effects are wavelength-dependent. This dependency must somehow be mediated by neurochemical channels connecting the photoreceptors with the endocrine system. And these could involve photoreceptors with no visibility function.

Dr. Alexander H. Friedman and Dr. Charles A. Walker of the Stritch School of Medicine of Loyola University in Chicago have reported circadian rhythms in rat mid-brain and caudate nucleus biogenic amine levels that respond to the light-dark cycle. They note that these amine rhythms can be altered to varying degrees by a change in the light-dark programming (1968). Dr. Robert Y. Moore, of the University of Chicago, working with Drs. Heller, Bhatnager, Wurtman and Axelrod, reported in a paper entitled *Central Control of the Pineal Gland: Visual Pathways (Archives Neurology,*Volume 18, pages 208-218, February 1968) that the results of their studies indicated that:

> These findings establish a separate function for the inferior accessory optic tract components of the central visual projections in the maintenance of light-mediated neuroendocrine responses.

And in the February 6, 1970, issue of *Science* (Volume 167, pages 884-885). Lennart Wetterberg, Edward Geller and Arthur Yuwiler of the Neurochemistry Laboratory, Veterans Administration Center and Department of Psychiatry, University of California School of Medicine, state that removal of the Harderian gland abolishes the response of the circadian rhythm of pineal serotonin to the influence of light in neonatal rats. The Harderian gland in deer was first described in 1694 by J. Harder. It is located around and behind the eye and is found in all vertebrates, with the exception of the higher primates:

> Its function is unknown, but speculation has ranged from a source of lubricant for the eye to gonadal regulation through merocrine secretion. The significance of a possible extraretinal photoreceptive function for the Harderian gland is still difficult to assess.

This article also mentions that Benoit has observed in the duck that hooding prevents light stimulation of the testes and that testicular stimulation occurs if light is presented to the lateral side of the head after section of the optic nerves, or if light is introduced into the eye socket after removal of orbital tissue. Since birds have well developed Harderian glands, these results could be due to stimulation of this gland.

Michigan State University endocrinologist, Dr. Joseph Meites, in 1969 stated that light entering the eyes causes nerve impulses that influence the lower brain and pituitary gland that trigger the release of other horrnones. Dr. Meites further states:

> We have no idea how many diseases are linked with hormone problems, but we do know that several diseases such as diabetes, infertility, cancer and thyroid disorders are involved with hormone imbalance.

During the past ten years Dr. Meites' laboratory has been active in studying neurohormonal mechanisms in the brain.

Now, getting back to the article I submitted to *Survey*, I quickly followed the suggestion of Dr. Leopold to send the material to the A.M.A. Shortly thereafter it was published in a nine-page story, including three pages of illustrations in color, by the A.M.A. in the March, 1963, issue of *Today's Health*, and at the end of the article, *My Ivory Cellar* was recommended as further reading on the subject. A short reference to the article was made in the *Journal of the American Medical Association* (August 3, 1963, 48: Med News), and in a news release from the public relations department of the A.M.A. Soon other articles appeared, mentioning the effects of different wavelengths of light through the eye, influencing tumor development in mice. Two in particular I

91

thought were exceptionally good—one in *The Chicago Tribune Sunday Magazine Section* of April 28, 1963, and the other in the *Kiwanis* magazine for December 1963/January 1964. The manuscripts for both of these articles were submitted to me for proofreading and then to both the A.M.A. and A.C.S. headquarters for final approval before being released. I specifically mention that both manuscripts were approved by the A.M.A. and A.C.S., a circumstance which more than alleviates some of the harsh criticisms made by others. I'm no stranger to criticism, and I recall one particular example, relayed to me in a letter dated May 16, 1961, by a friend, Hedwig S. Kuhn, M.D.; director of the department of ophthalmology at the Hammond Clinic, Hammond, Indiana:

Dear Hedie:

I have been working as closely as possible with our mutual friend, John Ott, and believe we have something underway at the University of Virginia with (Dr.) Ebbe Hoff which may settle some of the questions once and for all. But I will have to admit that if John doesn't quit curing cancer by shining a light in everyone's eye, I am not going to be able to accomplish anything for him....

Kindest personal regards. . .

> Most sincerely,
> (signed) "Chuck"
> Charles W. Shilling, M.D.
> Director, BSCP

Nine

ANIMAL RESPONSE TO LIGHT

All of my research, plus external information that came to me from many sources seemed to indicate, more and more, that animals respond to the intensity, periodicity and wavelength distribution of light in much the same way that plants do. Certainly there was sufficient additional evidence to warrant submitting another grant application to the National Institutes of Health. Several doctors prominent in research work wrote letters of recommendation to Dr. James Shannon, Director of The Public Health Service of The National Institutes of Health, resulting in his taking a personal interest in our effort and offering some guidelines to follow in submitting the next grant application. Six of the doctors, including several members of our Board of Directors, agreed to act as collaborators and actively participate in the project. They all helped in preparing the forms for the application.

The six doctors included Robert Alexander, M.D., Chief Pathologist, Presbyterian St. Luke's Hospital, Chicago; Samuel Lee Gabby, M.D., Senior Member of the staff and member of the research committee, Sherman Hospital, Elgin, Illinois; Elliot B. Hague, ophthalmologist and Chairman of the New York Academy of Sciences Conference on "Photo-Neuro-Endocrine Effects in Circadian Systems with Particular Reference to the Eye;" Irving H. Leopold, M.D., ophthalmologist, Director of Wills Eye Hospital, Philadelphia: Frank J. Orland, D.D.S., Director of the Walter G. Zoller Memorial Dental Clinic, of the University of Chicago, where we had previously assisted with an experiment that showed a relationship between the amount of tooth decay and the type of light environment that laboratory animals were kept under (mentioned in *Annals of Dentistry*, Vol. XXVII, No. 1, March 1968, Page II), and Edward F. Scanlon, M.D., member of the staff and Head of the Tumor Research Committee of the Evanston Hospital in Evanston, Illinois.

Meanwhile, I was invited to speak before the Sarasota County Medical Society and show the time-lapse pictures. Shortly thereafter, Dr. Thomas G. Dickinson, who had originally put me in touch with the Wills Eye Hospital, held a special meeting at his home and invited other doctors who had indicated their interest in our light experiments. One of the doctors was Roscoe Spencer, M.D., who had recently retired as Medical Director of the United States Public Health Service, where he had been in charge of cancer research. He wrote a strong letter of recommendation to the new Director of the National Cancer Institute, Dr. Kenneth M. Endicott, urging that favorable consideration be given to our currently pending grant application. A month later a letter advised us that the National Advisory General Medical Sciences Council did not recommend approval of our application. Upon our request for specific reasons, we were informed as follows:

> Our reviewers carefully examined your proposal to investigate biological responses to specific action spectra. They observe that no details of the experimental design are given; current literature on photobiology is casually mentioned, and pilot experiments are referred to but not adequately described. Finally, this vague proposal gives no evidence of a basis in scientific fact or method. They further commented that it is not known that the applicant has any specific background which would permit him to analyze physiological phenomena in a meaningful way, and that the six persons named as collaborators have impressive titles and affiliations, but

94

nothing is offered in support of their competence to participate in the project.

Chalk up one more disappointment and a big one this time. It was a real letdown to have to stop thinking about plans for an expanded research program and to get down to the realities of making the most of our existing limited facilities and finances.

Despite this handicap, we began to study the effect of different colored light environments on mice. It had been necessary to start quite a breeding colony of rabbits for the pigment epithelial cell experiment. There just wasn't any available space to accommodate the larger rabbit cages, so the only solution was to push aside some of the time-lapse projects involving plants to make way for the rabbit cages. I soon found that the results we were obtaining in breeding rabbits in the ultraviolet transmitting plastic greenhouse were so far superior to the average results obtained in artificially-lighted animal breeding rooms that the delay of the time-lapse sequences involving plant studies was more than offset.

I began thinking about the results of adding a little long wavelength or black light ultraviolet to the ordinary incandescent light source of the microscope and it seemed to me to be quite significant, especially when coupled with the exceptionally satisfactory results obtained in the breeding of rabbits in the ultraviolet transmitting greenhouse. This prompted me to build additional space where laboratory animals could be exposed to natural sunlight conditions through different types of glass and plastic materials that would allow various amounts of natural ultraviolet to penetrate. The additional space also contained a compartment with a simple air curtain that would allow the natural sunlight to penetrate unfiltered in any way. The opening was screened to keep insects out and the air was circulated from the center of the animal room through the various compartments and out the air curtain, in a way that would keep the cold weather out during winter.

The improved results in breeding not only the rabbits in the ultraviolet transmitting greenhouse as compared to the standard artificially-lighted breeding room, but also the differences noted in breeding both mice and rats under the air curtain as compared to the various colored light compartments were almost unbelievable. In breeding laboratory animals it is standard procedure to remove the male from the cage before the litter is born because of his tendency toward cannibalism. However, the male rats in cages exposed to sunlight through ultraviolet transmitting plastic, quartz glass or air curtain were observed to help care for the litter, especially when the female was removed from the cage. Furthermore, the adult male rats appeared decidedly more docile and friendly when handled, whereas those kept under fluorescent light seemed more irritable and developed a tendency to bite.

Although these results were interesting, the numbers of experimental animals did not approach statistical significance. However, later experiments involving larger numbers of animals did begin to show significantly different responses when kept under different colors or wavelengths of light.

In a group of 536 mice born under the air curtain, ultraviolet transmitting plastic or quartz glass, all except 15 survived to maturity. This represented a survival ratio of 97 per cent. Under various types of fluorescent light 88 per cent of 679 mice survived to maturity. Approximately 94 per cent survived to maturity under cool white, warm white and daylight white lamps, but the percentage of survival noticeably declined under the different, deeper-colored lights. The lowest survival rate of 61 percent occurred in the mice exposed to pink fluorescent light.

The animals being kept outdoors under natural sunlight coming through ordinary window glass, ultraviolet transmitting plastic, quartz glass and the air curtain, also showed significantly different responses. The tails of both male and female C_3H mice became spotted and would develop sores under 12 hours daily exposure for three months to pink fluorescent light. At this stage of development, when some of the mice were transferred back to the air curtain, the tails would again appear normal after thirty days. If, however, the mice were left under the pink fluorescent for six months, the tails would appear gangrenous and, bit by bit, slough off until some of the more severe conditions resulted in complete necrosis of the tail. A careful examination of this tail condition revealed no evidence of the presence of any bacteria or fungi.

A subsequent showing of the mice pictures brought the following letter from Phyllis A. Stephenson, M.D., who is now a member of our own Medical and Scientific Advisory Board.

March 2, 1970

Dear Dr. Ott:

As a practitioner of medical oncology, and having spent two years in research in tumor immunology, I have noted particular difficulty with two strains of mice at our laboratory at the Sloan-Kettering Institute. One is the C_3H mouse, the other the SJL/J mouse, as far as the tail lesions are concerned. Multiple attempts to find the etiology of this have been made at the Jackson Memorial Laboratory, our own particular laboratory, and others. No definitive cause has been found: no good article is in the literature. It has been felt that bacterial infection (streptococcus?), humidity and overcrowding does contribute to these particular lesions. However, placing the animals in cages with smaller numbers of others and giving them forms of activity such as a running wheel did not significantly lower the incidence of lesions. It is significant that these tail lesions alone have been known to decrease the nutrition of the mice sufficiently to eliminate them from significant laboratory experiments. Multiple microscopic sections and cultures for bacteria and fungus from the lesions of these mice find only an irritative phenomena in filtration of white cells and lymphocytes into the area affected, with no constant bacteria or fungus. Autopsy of these mice show no internal tumor process. The lesions become so bad that the mice lose their tails, with infection at the stump which occasionally invades the rectum, causing an obstructive type of situation. The mouse becomes completely debilitated and may die.

In my experience we have never used any different mode of lighting for its work-up. Most of the time these mice are sacrificed by the laboratory which bought them. We did consider the tail lesions a major problem in our laboratory. These two strains of mice are used commonly in research: the SJL/J for lymphoma and leukemia studies (TL 1,2,3, positive) and the C_3H for mammary tumor and Gross virus studies. It is interesting that other mice do not consistently develop this tail lesion. It may be specific for the inbreeding of these two strains. Specific treatments such as topical and systemic steroids, topical and systemic antibiotics did not help. The SJL/J mice are very active, nervous mice, which are known to occasionally attack each other. However, the C_3H Bittner is a relatively docile animal and we cannot correlate the temperament of the animal with the tail lesion. Thus, I feel that the tail lesion is a significant problem with mice in the laboratory setting, and any answer we could find would be extremely helpful.

> Very sincerely,
> Phyllis A. Stephenson, M.D.

We were trying to find some of the answers ourselves.

Three months exposure of the same C$_3$H strain of mouse during the daylight hours to a new purple, plant growth fluorescent tube would result in the animals losing much of their fur, and after six months under the purple light they would lose almost all of their fur and appear to be in a relatively unhealthy-looking condition. Autopsies performed on some of the animals indicated a normal, healthy condition of the heart tissue in all those from the air curtain compartment, whereas all those from the pink fluorescent light showed an excessive condition of calcium deposits known as *calcific myocarditis.* The tails of the mice under the pink fluorescent light and the fur of the mice under the purple light were of course exposed directly to the lights. The abnormal responses may therefore have been due to the direct exposure of the tails and the fur to the light, or possibly these results could have been due to the light entering the eyes and stimulating the retinal or oculo-endocrine system which may control the body chemistry. However, the heart tissue is not directly exposed to the light and these results must therefore be mediated in some indirect way, possibly through the eyes and pathway to the endocrine system.

In a simple preliminary experiment, the cholesterol level in the blood of mice kept under dark blue fluorescent light was found to be higher than in the blood of mice under red light. Many of the male mice under blue light became obese; this tendency was not noticed in the females.

These inconclusive observations are mentioned only to suggest the extent to which light might possibly be a variable in such research studies, and again, they emphasize the need for positive scientific control of all light sources used in the laboratory. Consideration might also be given to establishing some standard color for walls, ceiling and floors. In the October 25, 1963, issue of *Science*, the effects of isolation stress on white rats was reported. A marked difference in the toxicity of isoproterenol between community-caged rats and isolated rats provided a criterion for following the development of isolation stress. According to this group of investigators, the reversibility of isolation stress was also established, the reversal being attributed to the effects of short term versus long term isolation. Short term isolation stress was defined as one to ten days and long term from ten to thirty days. My particular interest in this article centered on the reported condition of *caudal dermatitis*, or scaly tail, which also followed a reversibility pattern attributed to the short term versus long term isolation stress. Although this condition was not nearly as serious as the condition of the tails of our mice under the pink fluorescent light, the similar reversibility patterns were definitely of interest. Of course the condition of complete necrosis of the tail could not be reversed, but during the early stages of development it seemed quite definite that the reversibility was attributable to the mice being transferred from the pink fluorescent to the full natural sunlight received through the air curtain.

My interview with the principal investigator in charge of the isolation stress study, and all of his co-workers, revealed that coincidentally, at the precise time between that designated as short term and long term isolation stress, the lighting environment of the two groups of animals was drastically changed. All the cages containing the isolated animals were on one large rack at one end of the animal room, with the cage doors directly facing the windows. The colony groups of animals were on another similar rack in a dark area and with the cage doors facing away from the windows. During the interview, it was learned that one of the laboratory assistants had moved the racks when mopping the floor and had inadvertently switched their locations. Needless to say, this emphasizes the need for having laboratory light sources under scientific control.

Continuing with our experiments with the C_3H strain of mouse, which is so highly susceptible to spontaneous tumor development, our purpose was to carry out further experiments to determine the length of time required for such tumor development in the animals kept under different light environments. This part of the experiments did, I believe, reveal possibly the most significant data of all of our experimental work. The C_3H strain of mice kept under pink fluorescent light developed spontaneous tumors and died, on the average, in 7 1/2 months. The animals under different types of light with an increasingly wider spectrum showed a progression in life span up to 16 1/10 months. Over 2,000 mice were used in this experiment.

It is interesting to note that the most extreme adverse conditions regarding not only tumor development but also necrosis of the tails, calcium deposits in

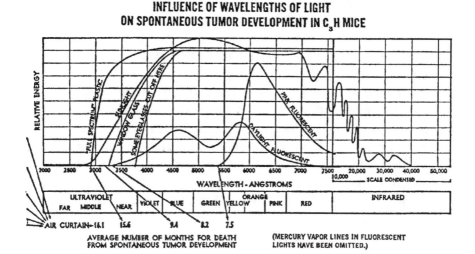

INFLUENCE OF WAVELENGTHS OF LIGHT
ON SPONTANEOUS TUMOR DEVELOPMENT IN C$_3$H MICE

AVERAGE NUMBER OF MONTHS FOR DEATH
FROM SPONTANEOUS TUMOR DEVELOPMENT

(MERCURY VAPOR LINES IN FLUORESCENT
LIGHTS HAVE BEEN OMITTED.)

101

the heart tissues, smaller numbers in litters, and diffi-cult behavioral problems, all were caused by pink light.

The question then arises that if these responses are due to the absence of the shorter wavelengths and the influence of the longer wavelengths, why then are some of these responses not most severe under red light—as in the case of the chloroplasts in the cells of Elodea grass? The answer to this is not clear, but considera-tion might be given to the fact that many nocturnal animals do not see red light because these wavelengths are beyond the range to which their visual receptor mechanism is responsive. Many zoos are now using red light in rooms where nocturnal animals are located, and the animals seem to think it is nighttime and are accordingly more active and interesting to watch. When ordinary lights are turned on the animals usually curl up and go to sleep.

The indication that some nocturnal animals cannot see red light, and the suggestion that other nocturnal animals can "see in the dark" because their eyes are sensitive to infrared is somewhat contradictory.

However, the range of wavelengths to which the vis-ual receptors of nocturnal animals respond may vary with different species. Likewise, the usual receptors may not respond to the precise same range of the longer wave lengths that activate the oculo-endocrine system, which seems to be definitely so with the ultraviolet.

The results of the spontaneous tumor development experiment under different types of lights were indeed quite startling, and several doctors interested in cancer research who had assisted in setting up the proper controls for our experiments agreed that this experi-ment should certainly be repeated, and several agreed to do so at the various hospital or research laboratories with which they were associated.

Dr. Samuel L. Gabby, who had conducted the experiment with the mice in his basement, offered to keep some outside as well if we could make him some sort of portable enclosure with an ultraviolet transmitting cover. We did, and he obtained virtually the same results as shown in the tumor development chart.

I was also asked to show the films and speak at one of the research seminars at The Evanston Hospital, north of Chicago, where Dr. Edward F. Scanlon, one of our trustees, was head of the Tumor Research Committee and had been named as principal investigator in our last N.I.H. grant application. Immediately following my presentation, Dr. Scanlon advised that work was actually in progress in injecting hamsters with several different tumor transplants in connection with their studies of the effectiveness of various anti-tumor drugs. He suggested that they would postpone the injection of the drugs for the time being if I would take half of the animals, selected at random, and keep them in the air curtain compartment in our laboratory animal quarters where they would be subjected to natural daylight. The other half would remain in their regular laboratory quarters under cool white fluorescent tubes. This plan was agreed on.

Dr. Scanlon and several members of his staff periodically visited our laboratory as the project proceeded. The results of this experiment showed that the animals that had received a very fast acting tumor transplant showed little difference in the life span from those under the air curtain and those that remained in the regular animal laboratory facilities at the hospital under cool white fluorescent tubes. However, those animals that received a slower acting type tumor transplant did show a significant difference. The animals remaining under the cool white fluorescent tubes showed an average life span of 29 days, whereas those kept under the air curtain averaged 43 days.

With the unanimous approval of the entire research committee of the hospital, Dr. Scanlon wrote a report, together with a request for a small research grant to carry on further studies, and submitted this to the Illinois branch of the American Cancer Society. The Illinois branch in turn forwarded the report and request on to the American Cancer Society headquarters in New York City, and in due course the following reply was received from the Illinois office:

> In a memorandum under date of December 9 we have received information from the Assistant Vice President for Research at National that the Advisory Committee has recommended disapproval of this application, with the following comments:
>
> It is proposed to study the effects of visible and near-visible light on the growth of transplantable hamster cancer in hamsters, by exposing inoculated animals to various bands of the spectrum and following the animals through their life spans.
>
> It should be noted that the assay—life span—is a summation of a host of factors not necessarily connected directly with tumor behavior *per se*. The results will be difficult or impossible to interpret in any meaningful way. Any direct effect of light on tumor cells cannot be observed in this study and no evidence exists or is presented to warrant the belief that such exists. No account is taken of the penetration of visible and near-visible radiations into the animal or the tumor cells, nor of thermal and photochemical effects (e.g. burns); statistical precision for meaningful correlations will be insufficient.
>
> While there is every likelihood that exposure to different kinds of light will affect certain physiological response in the animals, they (sic) will only confuse the issue.
>
> Support of this proposal and project as presented is not justified on scientific grounds.

The power and authority of such a distinguished

scientific committee is awe inspiring. There is no recourse. Their word is final. Their combined knowledge and wisdom is supreme. But how can anyone be so certain as to what *cannot* cause cancer until it is known what *does?* To say that considering light as a variable would only further confuse the issue is difficult to reconcile with the basic concepts of research.

Being turned down again was discouraging, to say the least, but none of the reasons yet given by any reviewing committee has offered any convincing evidence that light energy might not be a missing link. In fact, all of the reasons given opposing the suggested light hypothesis indicated to me a great need for additional and determined pursuit of the subject.

Ten

BIOLOGICAL EFFECTS
OF TINTED LENSES

The existing confusion (with regard to the effects of various tints) applies not only to matters relating to visible light, but seems to become even greater with regard to ultraviolet.

The fact that too much heat will produce a severe burn, and a little extra oxygen in the incubators of premature babies can cause blindness known as *retrolental fibroplasia*, does not necessarily mean that an environment of absolute zero temperature, totally devoid of oxygen, is indicated as desirable. However, this obviously irrational conclusion is being generally applied because of the known harmful effects of excessive exposure to ultraviolet.

The full spectrum of natural sunlight, including both the visible and invisible rays, under which life on this earth has evolved, is known to be of direct benefit to man. In reporting on the effects of exposure to ultraviolet, Dr. Ellinger, in his book *Medical Radiation Biology* (Charles C. Thomas) writes:

> Irradiation of human subjects with erythema-producing doses of ultraviolet results in an improvement of work output. In studies on the bicycle ergometer, it has been shown that under these laboratory conditions the work output could be increased up to 60%. Analysis of this phenomenon revealed that the increased output is due to decreased fatigability and increased efficiency. Cardiovascular responses served as an indicator.

In 1967, at the meeting of the International Committee on Illumination (C.I.E.) in Washington, D.C., a

106

paper by three Russian scientists, Dantsig, Lazarev and Sokolov, was presented which stated that:

> If human skin is not exposed to solar radiation (direct or scattered) for long periods of time, disturbances will occur in the physiological equilibrium of the human system. The result will be functional disorders of the nervous system and a vitamin-D deficiency, a weakening of the body's defenses, and an aggravation of chronic diseases. Sunlight deficiency is observed more particularly in persons living in the polar regions and in those working underground or in windowless industrial buildings.
>
> The simplest and at the same time the most effective measure for the prevention of this deficiency is the irradiation of human beings by means of ultraviolet lamps. Such irradiation is conducted either in special rooms called photaria or directly in locations where persons are regularly present—in workshops, schools, hospitals, etc. As a rule, the daily dosage of ultraviolet does not exceed half of the average dose which produces a just perceptible reddening of an untanned human skin. It is preferable to use fluorescent lamps which use phosphor and have a maximum emission of 315 nm.* The beneficial effect of ultraviolet irradiation has been confirmed by many years' experience.
>
> Ultraviolet irradiation is also beneficial for agricultural animals.

An unnecessary degree of fear of ultraviolet exists, probably as a result of a general lack of understanding of the difference in the relative intensities of the near, or long, wavelength ultraviolet and the far, short-wave ultraviolet in natural sunlight at the surface of the earth. The atmosphere filters or stops virtually all of the far short wavelength ultraviolet except for a trace amount, but does allow the near long wavelength

* This is in the long wavelengths black light range of ultraviolet.

ultraviolet to pass through in amounts comparable to the intensity of visible light. Thus, life on earth has evolved under the balance of short wavelength ultraviolet comparable to the very low levels of general background radiation and much higher intensities of long wavelength ultraviolet comparable to that of visible outdoor natural sunlight.

Many artificial "sun" lamps manufactured today give off a peak of energy in the far short wavelength ultraviolet that is filtered from natural sunlight by the atmosphere. They are the same as germicidal lights, and these can produce severe burns and injury. This type of ultraviolet light has been used extensively in clinical experimental work and has shown beyond any doubt that over-exposure will produce harmful results, including skin cancer in laboratory animals.

The question then arises on how long an exposure and at what intensity, constitutes over-exposure. In view of the apparently extremely delicate biological responses to minor variations in energy levels in nature, it would seem that not very much of an increase of intensity of short wavelength ultraviolet over the trace amount in natural sunlight would be necessary to upset nature's biological balances. An actual measurement of the trace amount of short wavelength ultraviolet in sunlight is difficult to establish. The spectral energy chart of sunlight published by the U. S. Bureau of Standards totally ignores it, and shows an absolute cut-off in the ultraviolet range at approximately 2900A, as a result of the filtering effect of the atmosphere. My spectral chart shows the line representing sunlight energy continuing from 2900A on into the shorter wavelength at the very bottom of the chart in order to represent this trace amount of far, short-wave ultraviolet in a pictorial way. This trace amount of short wavelength ultraviolet might be compared to the so-called trace amounts in chemistry, which at one time were totally ignored but are now recognized as being of very great importance, especially in biochemistry. Yet many scientists seem to feel a sense of accomplishment in being able to direct a high intensity microbeam of short wavelength ultraviolet on a small part of a living cell and then studying the abnormal growth responses, which may frequently be the ultimate death of the cell.

In a chapter on *The Absorption of Radiant Energy by the Ocular Tissues* in Duke-Elder's *Textbook of Ophthalmology* (C. V. Mosby), it is stated that "the thermal lesion caused by infrared rays is frankly pathological... The chemical or abiotic lesion [caused by ultraviolet rays], on the other hand, is of a completely different nature. Since the reaction is directly dependent on the absorption of energy, a critical threshold of wavelength and of intensity of radiation must be employed to excite it. A certain amount of abiotic activity may be evident at 3,650 (Coblentz and Fulton, 1924), or 3,500 A. (Newcomer, 1917), if conditions are favorable and the exposure sufficiently intense; it is more readily seen at 3,050 (Hertel, 1903; and Henri, 1912), but it is found that for practical purposes only, rays below 3,000 may be considered abiotically active, and these must be used in an intensity of about 2,000,000 erg-seconds per square centimetre (Verhoeff and Bell, 1916; Duke-Elder, 1926)."

The text further states that:

> Clinically, the Keratitis produced in this way, together with an associated conjunctivitis, produces the condition of photophthalmia, which occurs after undue exposure to the sun's rays (solar photophthalmia, snow blindness, etc.) or to artificial sources rich in short-waved light (industrial photophthalmia electrica, etc.)

Roughly 2,000,000 ergs is the equivalent of 19

109

minutes of full summer noon-day sunlight at Washington, D. C. Looking directly into the sun continuously for this amount of time would undoubtedly constitute over-exposure, even though almost all of the shortwave ultraviolet from sunlight is stopped by the atmosphere. However, 19 minutes of similar exposure to the equivalent intensity of an artificial light source rich in the shortwave ultraviolet is what is indicated as necessary to cause such abiotic lesions.

A paper presented by Dr. Frederic Urbach, et al, at a symposium held at the University of Oregon Medical School in 1965 states in the introduction that:

It has been suggested that prolonged exposure to sunlight may result in the development of skin cancer in man (Blum, 1959). As a result of the studies of Unna (1894), Dubreulh (1907), and many others (Blum, 1941), a number of arguments support the belief that sunlight is a causal factor in human skin cancer.

However, the following statement is included in the summary of the paper:

Squamous cell carcinoma of the head and neck were almost exclusively noted only on those areas which received maximal ultraviolet radiation while more than one-third of all basal cell carcinomas occurred on areas receiving less than 20% of the maximum possible ultraviolet dose. This suggests that some factor in addition to ultraviolet radiation plays a significant role in the genesis of basal cell carcinoma.

Certain ailments of the eye have also been related to excessive exposure to the ultraviolet in sunlight, and (as noted earlier) the practice of wearing sunglasses is becoming increasingly prevalent. It would be difficult to find an optician today who did not sell one brand or another of eyeglasses designed to filter out this so-

another of eyeglasses designed to filter out this so-called "harmful" ultraviolet radiation and prevent it from entering the eyes. Yet the paradox of this theory about the harmful effects of ultraviolet from sunlight is that scientific studies relating a high rate of pterygium, an abnormal growth on the eyeball that destroys vision through exposure to high intensity sunlight in the tropics, did not take into consideration whether or not those people with pterygium wore any kind of eyeglasses or sunglasses which would protect the eye from the ultraviolet part of the sunlight spectrum. Even ordinary eyeglasses filter out much of the ultraviolet in sunlight.

One extensive study of this subject gave as a major exception to these findings a group of Cree Indians in northern Manitoba, Canada, who had an exceptionally high rate of pterygium, and this far north they would definitely be out of high intensity tropical sunlight. A personal investigation of the situation revealed that this same group of Cree Indians had been issued specially designed sunglasses, of the wraparound kind, trimmed with leather to prevent even the slightest bit of unfiltered sunlight from reaching the eyes, in connection with an earlier experiment designed to study problems of glare, etc., from the snow and ice.

Neither study indicated whether the rate of pterygium was greater in the cases of those wearing sunglasses or not. It would seem, however, that this question might be pertinent, and in view of the combined overall results of both experiments, might raise the question as to whether the high incidence of pterygium resulted from actual direct exposure of the eyes to high intensity sunlight, or might possibly indicate the need of further studies to determine if various unhealthy conditions of the eyes could result from being deprived of the complete spectrum, including the normal amount of ultraviolet in natural sunlight which might be essential to maintaining a healthy condition. In checking a limited number of individuals who had developed pterygium while on military duty in the tropics, it was found that all had constantly worn prescription sunglasses.

In studying the harmful effects of ultraviolet, it has been common practice to consider only the effects on the part of the skin or eye that has been directly exposed to the sunlight. However, the more recent knowledge of the existence of an oculo-endocrine system greatly expands the research possibilities of the effects of ultraviolet, or especially the lack of it, on the retinal-hypothalamic-endocrine system.

Could the lack of the normal amount of ultraviolet in sunlight received through the eyes possibly cause a condition of hormonal or chemical imbalance and in turn make the skin hyper-sensitive to sunlight as far as skin cancer is concerned? It is known that some drugs and certain ingredients in soaps and cosmetics make people more sensitive to light. The question of any possible connection between different conditions of light sensitivity and hormonal imbalance or malfunction of the endocrine system might well be worth further investigation.

The amount of light actually entering the eye depends on the size of the pupil which is controlled by the iris. Under bright light, the pupil is normally much smaller so that only a fraction of 10,000 foot candles of full sunlight gets through to the retina. The pupil enlarges to let in proportionally more light of lower intensities. Thus the iris compensates to a great extent for such extreme variations in the intensity of the light entering the eyes but does not alter its wavelength distribution.

Tanning of the skin accomplishes essentially the same purpose of cutting down the intensity of the light that penetrates the outer surface and in this way helps prevent sunburn. The leaves of a plant respond in much the same way and can also be severely sunburned and will die if moved too suddenly from a shady location into full sunlight. Many species of plants that normally grow in shady locations never can fully adapt to bright sunlight.

In an article entitled *Degeneration of the British Beef Breeds in the Tropics and Subtropics*, by Jan C. Bonsma, *Breeding Beef Cattle for Unfavorable Environments*, p. 19, it is stated:

> ...pigmentation of hide is of the utmost importance to the breeder of cattle in tropical and subtropical regions. Ultraviolet radiation sets up irritation in the hides of cattle which lack pigmentation, causing hyperkeratosis. Lack of pigment in and around the eye makes... animals vulnerable to conditions such as eye cancer.

In 1969 an interesting experiment was conducted by Philip Salvatori, F.I.A.O. Mr. Salvatori is chairman of the Board of Directors of Obrig Laboratories, one of the largest manufacturers of contact lenses. He is also one of the trustees of the Environmental Health and Light Research Institute. The experiment consisted of fitting a patient with an ultraviolet transmitting contact lens for one eye and a non-ultraviolet transmitting lens over the other eye. Indoors, under artificial light containing no ultraviolet, the size of both pupils appeared the same, but outdoors, under natural sunlight, there was a marked difference. The pupil covered with the ultraviolet transmitting lens was considerably smaller. This would seem to indicate that the photoreceptor mechanism that controls the opening and closing of the iris responds to ultraviolet wavelengths as well as visible light.

When the ultraviolet wavelengths are blocked from entering the eye, the pupil remains larger than it would otherwise normally be and the visible part of the spectrum would then seem brighter. This could explain why some people feel a greater need for dark glasses.

The consistently better responses in all the experiments with both plants and animals to the full spectrum of natural sunlight, including its normal intensities of ultraviolet, and the effects on both the chloroplasts and the pigment granules when a little ultraviolet was added to the ordinary incandescent light source of the phase-contrast microscope, started me thinking about the possibility of adding some black light ultraviolet to the light sources being used in the various compartments where some of the animals were being studied. As previously mentioned, the intensity of the ultraviolet used in the microscope experiments was arrived at through trial and error, and too much ultraviolet was found to kill the cells in the microscope slides.

As I did not want to give the living animals too much ultraviolet to start with, I was not certain just what intensity would be within a safe limit. While in process of trying to decide how much ultraviolet to give the animals, my wife and I had dinner one evening in a restaurant known as "Well of the Sea," in the basement of the Hotel Sherman in Chicago. As soon as we entered the restaurant I noticed that there were black light ultraviolet lights placed throughout the ceiling. They had been installed solely for ornamental purposes, to cause designs on the waiters' coats, as well as the menus, to fluoresce in the otherwise subdued light. The next morning I went back to the restaurant with a meter to measure the intensity of the ultraviolet at various distances from the ceiling, such as table level and the eye level of the waiters as they walked directly under the various light fixtures.

I also wanted to ask the captain of the waiters a number of questions. In view of the general concern, especially at that time, regarding danger of over-exposure to ultraviolet, I wondered how long the lights had been installed and whether he had experienced an unusually high turnover among the personnel working in the restaurant. I asked him if any of his men complained of any eye problem, skin cancer, or other difficulties, such as sterility, which might be attributable to working for long periods of time under the black light ultraviolet. The captain told me that he had essentially the same group of men working for him as he had when they had opened the restaurant 18 years before. He said that the ultraviolet lights had been in use continually during that time, and that the health record of his men had been so consistently excellent that the manager of the hotel had checked into the situation, with medical supervision, to try to determine why this particular group of men was always on the job, even during flu epidemics, when other departments in the hotel would be short-handed because of employees' illness.

I then talked to the manager of the hotel, who told me that these men working in the "Well of the Sea" seemed to be a particularly happy group—courteous and efficient, and all seemed to get along well together. He said no explanation had been found to explain this, and that, at the conclusion of the study, it was thought to be simply a coincidence that this particular group of men should be so healthy and content. I asked if the men had been given a health check-up at the time they were hired. The manager explained that this was not customary and that the men just happened to be at the head of the list when the waiters' union was called on to staff the restaurant. I went back again on several occasions to talk with the captain and his men, and also to check to see if any of them were wearing glasses that would block the ultraviolet from entering the eyes. Not one of them wore glasses, which is rather unusual in this day and age, and none had ever complained of any eye problems or discomfort as the result of the ultraviolet light. Therefore, the measurements I was able to make of the intensity of the ultraviolet at the "Well of the Sea" gave me a good clue as to a safe level of exposure to start with for the laboratory animals.

Several months later at the Seaquarium in Miami, Florida, I noticed a similar black light ultraviolet light over some of the fish aquariums. In discussing this with the curator, Dr. Warren Zeiller, his assistant, Mr. Bevan, and some of their staff, I learned that these lights had originally been placed over some of the aquariums for decorative purposes, to give the fish an eerie but attractive appearance. Dr. Zeiller told me that the added black light seemed to solve one of their main problems in keeping fish. This was a condition of *exophthalmus*, or pop-eye, recently identified as due to a virus. I was told that it is rare that any aquarium fish are troubled with *exophthalmus* when kept in an outdoor aquarium under natural daylight and nighttime conditions. Another problem of fin-nipping also disappeared under natural conditions. Dr. Zeiller and Mr. Bevan have since written a number of articles on this subject, and report that certain fish that could never before be kept in captivity thrive under this added black light ultraviolet. Similar reports have been received regarding reptiles, birds, and animals kept in a number of zoos throughout the country.

One report on reptiles came from Jozsef Laszlo of the Reptile Department at the Houston Zoological Gardens in Texas, and appeared in the 1969 *International Zoo Year Book*, published by the Zoological Society of London. Dr. Laszlo reported that a number of reptiles and amphibians became noticeably more active when the cool white and daylight white fluorescent tubes in their cages were replaced with full-spectrum lighting. He further mentioned that it was even more interesting to see that some long starving but otherwise healthy snakes accepted food only a few days after the new lights were installed. One very rare snake of a type notoriously difficult to keep alive for any length of time in captivity ate for the first time since arrival in the zoo six months earlier.

At the Bronx Zoo in New York City, according to an article in the November, 1971, issue of the *American Cage-Bird Magazine,* it took four years for the curator to find out how to make the tufted puffin feel at home. Although the shy sea birds' northern habitat had been faithfully duplicated—rocky-cliffs and a consistently cool temperature—the birds refused to breed. With the installation of a new full-spectrum lighting system, the puffins have since attained a more natural coloration and for the first time in captivity one pair produced a fertile egg.

Another noteworthy item comes from Syracuse, New York, where Charles T. Clift, Director of the Burnett Park Zoo, reports that new lights installed in an attempt to stop vandalism fooled many of the animals into thinking that spring had arrived. "The zoo has been turned into a veritable maternity ward. The cougars fell in love all over again and produced their fourth litter. We collected five goose eggs. At least eight lambs were born, and the deer population increased by twenty. Big Lizie gave birth to a bear cub. The wallaby produced a new mini-kangaroo and the chimpanzee is expecting in August."

A significant difference in the amount of voluntary activity in mice kept under different colored lights was reported in the April, 1969, issue of *Laboratory Animal Care* by J. F. Spalding. The activity was measured by the number of revolutions of a rotating activity wheel in which the mice were free to run.

All the mice tested, regardless of sex, age or color, exhibited activity related to six different color-environments, as follows: Group 1, red and dark; Group 2, yellow; and Group 3, blue, green and daylight. The groupings are given in order of their activity, with Group 1 showing the greatest. This response is interesting to compare with the results obtained in the zoos using red-lighted "night" rooms.

Dr. Spalding mentions that white albino mice responded to environmental lighting changes to a greater degree than black mice, and that there were further differences in activity due to age and sex. Of particular interest were findings reported of an earlier experiment indicating that different lighting conditions in the visible color spectrum had a strong influence on activity in normal mice, but that enucleated mice showed equal activity in the dark and in all color environments. Dr. Spalding further suggests that the results of these experiments may be pertinent to environmental lighting conditions not only of stock animals but also of the working man.

During the winter of 1968-1969 a serious outbreak of Hong Kong flu swept the country. Florida was no exception. The Health Department reported 5 per cent of Sarasota County—or 6,000 people—sick with the flu at one time. Employee illness caused the temporary closing of one supermarket, a social club, and the shutdown of two areas of the Sarasota Memorial Hospital because sixty-one nurses were out with the flu.

Obrig Laboratories, located just north of Sarasota, is one of the largest manufacturers of contact lenses and has approximately one hundred employees. During the entire flu epidemic not one employee was absent because of any flu type ailment, according to Philip

Salvatori, Chairman of the Board.

Obrig Laboratories was the first to design a new building using full-spectrum lighting and ultraviolet-transmitting plastic window panes throughout the entire office and factory areas. The added ultraviolet seemed to tie in closely with the results noted at the "Well of the Sea" restaurant in Chicago. Mr. Salvatori also mentioned that the Obrig employees had not been given any mass inoculation against the Hong Kong flu, although some individuals may have received shots from their private physicians. Mr. Salvatori also commented that everyone seemed happier and in better spirits under the new lighting, and that work production had increased by at least 25 per cent.

On another trip to Florida I gave a lecture to an advertising club, and after I had finished my talk, Mr. Richard L. Marsh, manager of radio station WILZ near St. Petersburg, told me of a similar situation. He said that some of the staff at the radio station had taken it upon themselves to try to brighten up their surroundings in both the studios and the control rooms by replacing the regular white fluorescent tubes with those of a deep pink color. About two months later, they began to have personnel problems. For example, announcers began performing poorly on the air. Everyone became irritable and consistently at odds with management decisions and generally difficult to control. Two resignations were received from employees without any known reason for their wishing to leave other than general dissatisfaction with themselves and the staff.

Then, one morning one of the men said, "If those pink bulbs aren't removed I'll go out of my mind." That sparked an immediate reaction, and that very day all of the pink tubes were removed and replaced with the white tubes. Within a week, as if by a miracle, tempers ceased to flare, congeniality and a spirit of working together began to redevelop and resignations were withdrawn. The airwork improved, with mistakes at a minimum.

These results seemed quite in line with the preliminary reports I had received from an experiment that I helped design to study the effects on mink kept behind different colored glass and plastic.

The experimental work with mink was carried on at the Northwood Mink Farms in Cary, Illinois, but unfortunately the project was suddenly interrupted due to the death by automobile accident of Mr. Bud Grosse, owner and operator of the farm. Immediately after his death the principal investigator and his two assistants all moved to other mink ranches in different parts of the country and no official paper was ever published. However, I was in close contact with Mr. Grosse while the experiment was under way and progress reports were given to me on the various results obtained.

The reports indicated that the mink exposed to natural daylight through a deep-pink glass became increasingly aggressive, difficult to manage and in many instances actually vicious. Ordinarily, mink are kept in open sheds with open window areas containing no glass. They are provided with a box-like shelter containing some straw, but the sheds are not heated as the natural habitat of mink is in north country, where the winters are long and cold.

However, mink normally are quite fierce and even without the pink glass it is customary for the animal caretakers to wear heavy leather gloves for protection, especially during the mating season. But when some of the mink were placed behind deep blue plastic they became friendly and docile, and in thirty days could be handled with bare hands like ordinary house pets.

The effect of the different colors on the animals'

120

behavioral patterns was interesting, but the difference in the results of mating the animals under either pink glass or blue plastic was possibly of even greater interest.

When a female mink does not become pregnant after the first mating, it is common practice to give her an injection of a pregnant mare serum before attempting the second mating. This was not necessary with any of the female mink in the cages with the blue plastic, as all became pregnant after the first mating. Furthermore, to use the language of the mink industry, all the males were found to be "working males."

But the situation was quite different with both males and females in the cages behind pink glass. After three attempts at mating the females, which included two injections of the pregnant mare serum, only 87 per cent became pregnant and 90 per cent of the males were classified as "non-working."

The principal investigator of the project was Alex Ott (no relation), who also advised that four animals under the pink glass died during the experiment from a strange malady that he had never seen before. An autopsy of each animal indicated what appeared to be a cancerous condition of the abdominal area including a number of vital organs. Unfortunately, an actual biopsy was not performed due to the abrupt termination of the entire project. Approximately 500 female mink were used in each experiment.

Another interesting bit of information turned up as the result of a questionnaire given to a group of college students by a professor of psychology. In general, the questionnaire asked if the students wore glasses or contact lenses, and—in particular—if they wore tinted contact lenses or sunglasses and if so, what color. Questions were included asking how much time was spent out of doors and how many hours spent watching television. Roughly 300 students answered the questionnaire, but the overall replies clearly showed that either more detailed questions would be necessary or, better, a personal interview.

However, one rather clear relationship did show up. Although not statistically significant because only three cases were involved, three students did reply that they constantly wore "Hot Pink" glasses and a check with the faculty ratings indicated that these same three students also were considered to be the most psychologically disturbed students in the college.

More recently the following interesting letter was received:

Dear John:
Thirty days have now passed since we changed one of our player's glasses from a pink-tinted to a medium gray as per your recommendation.

It was amazing to observe how the player was changed from a hyper-aggressive and helmet-throwing player to a very relaxed, confident person. There was a great deal of improvement in performance.

The performance of one of our other players who had mysteriously retrogressed for no evident reason has, since the removal of psychedelic-type red lighting from his dormitory room, regained his usual good performance.

Our entire staff would again like to thank you for the time you have spent enlightening us in areas that for too long have been explained only in vague generalities.
Sincerely,
Kansas City Royals Baseball Academy
Syd Thrift, Director

Eleven

EFFECTS OF RADIATION
ON BIOLOGICAL CLOCKS

So far, all of the growth responses observed in plants and animals as a result of the light study projects have had to do first with the visible portion of the light spectrum, and then with the addition of the adjacent ultraviolet wavelengths. This portion of the visible and UV light, though, represents only a small part of the total electromagnetic spectrum. There are many shorter and longer wavelengths that are capable of penetrating most types of building materials as readily as visible light penetrates ordinary window glass. Though similar in nature to visible light, some of these wavelengths are frequently referred to as general background radiation.

In thinking about various photobiological responses in plants and their flowers, it has become apparent to me that some of the responses, such as the difficulty I encountered with the morning glories, were reactions to different bands of wavelengths within the visible spectrum, and the problem of the apple not ripening was quite obviously from the lack of certain wavelengths within the ultraviolet part of the spectrum. But I noted still other responses in certain plants that did not seem to respond to either the visible or ultraviolet light, or even to variations in temperature. For example, the night blooming cereus, which belongs to the cactus family, ordinarily blooms in the evening as it gets dark, and then the blooms collapse as the sun rises the following morning. It blooms only during a normal nighttime period, whether the plant is indoors or outdoors, and cannot be forced into bloom during the daytime, even though placed in a totally dark

123

closet. The blossoms will collapse the following morning at the time of sunrise even if the plant remains in the dark closet. Other types of day blooming cacti that normally open during the daytime and close at night follow the same rhythm when placed in a dark closet.

Turning an ordinary incandescent light on and off in the dark closet during either the normal daytime or nighttime hours has no effect on either of these night blooming or day blooming cacti. This phenomenon became apparent when I realized that the photographic lights necessary for taking time-lapse pictures did not disturb the night blooming cereus or prevent the blossoms from opening during the normal nighttime period. It is interesting to place a night blooming cereus and a day blooming cactus side by side in a totally dark closet and note the blooms of the nocturnal plant open in the nighttime and then close during the daytime, while the diurnal plant responds oppositely, regardless of whether an ordinary incandescent light is on or off.

Another good example is the way a sensitive-plant folds its leaves and lowers its branches at night and then resumes its daytime position as the sun rises the next morning. If the sensitive-plant is placed in a dark closet during the daytime, the leaves remain open in daytime position until the sun sets outdoors. They will resume their daytime position when the sun rises even though they remain in the dark closet. These responses in plants that seem to become established in a normal light-dark cycle, but continue this established rhythm even though the light-dark cycle is altered, are called circadian rhythms and are attributed to a so-called biological clock system. Just how this mechanism works and where it is located in the plant has not yet been fully explained, but much is being written about it.

However, as far back as 1729, M. DeMairan submitted a paper entitled *Biological Observation* to the French

Royal Academy. He noted that the sensitive-plant folded its leaves at sunset in a fashion similar to the way in which the plant reacts to touch or agitation. He further noted that this phenomenon occurs even if the plant is kept in the dark and not exposed to the sun or the great outdoors. Mairan concluded that the sensitive-plant does "sense" the sun without seeing it in any way. Mairan commented on the possible relationship between this phenomenon and the unfortunate delicacy of a large number of patients who respond in their beds to the difference between the day and night outdoors.

In an attempt to explain the persistent rhythms of the sensitive-plant in the darkness, Mairan went on to suggest that all such rhythms are being forced on the organism by some unknown factor in the universe. Nevertheless, research procedures today, in studying the phenomenon of the so-called built-in biological time clock in relation to light and darkness, consider light only as that part of the total electromagnetic spectrum to which the human eye is sensitive. What the human eye does not see is generally thought of as darkness, with the connotation that no further radiant energy exists that could produce a photobiological or photosynthetic response. This may raise questions about similar responses that have been observed in so-called darkness and have been called chemosynthetic because of the theoretical absence of any light energy.

Accordingly, I designed an experiment to determine if some of the wavelengths of general background radiation, beyond the range of human vision, might be directly controlling at least some of the so-called circadian rhythms. Six sensitive-plants (mimosa pudica) were placed at noon in a dark closet at the basement level of a three story residential building. The door to the closet was made of wood but the walls and ceilings were of concrete, approximately four to six inches in thickness. The outer walls of the building were of brick and the roof of slate; interior construction was of wood and plaster.

The leaves of the plant remained open and the leaf stems, or petioles, in the upward daytime position until sunset. Then the leaves folded and the petioles dropped downward to the normal nighttime position. They remained in this state until sunrise, when both the leaves and the petioles resumed their normal daytime, open and upward, position.

As the only practical shielding against some of the general background radiation—especially cosmic radiation—is a massive amount of earth, six sensitive-plants were taken at noon to the bottom of a coal mine, 650 feet below the surface. The leaves and petioles of all six plants immediately assumed their nighttime position, not waiting for the sun to set. The area where the plants were placed was lighted with regular incandescent bulbs. This suggests that the day-night responses of the leaves and the petioles of the sensitive-plant react to some form of radiation capable of penetrating through the building material surrounding the "dark" closet at the surface of the earth, but not to the bottom of a coal mine, 650 feet down. This also suggests that these particular responses are not influenced by the wavelengths of light energy produced by an ordinary incandescent light bulb.

Another experiment confirmed what might be expected—that the degree of loss of response to the general background radiation was proportionate to the amount of shielding material involved.

The location of the sensitive-plants in the mine was approximately 100 feet from the bottom of the main elevator shaft which also served as the fresh air intake system. This was to eliminate the possibility of the presence of any form of coal gas that could affect the responses of the plants.

It was interesting to observe that after being down in the mine all night, the leaves of the plants would open just a little the following morning. Then, when they were brought to the surface, the leaves would open fully in the sunlight, but they would not respond to being touched, nor close in their normal way. This suggests that they did not get "charged up" with some form of nighttime radiation while down in the mine, and also raises questions about a possible interaction between daytime and nighttime radiations similar to that of the responses of phytochrome, a chemical within the cells of leaves that interacts and changes its form when exposed to different wavelengths within the spectrum of visible light.

The buds of the hoya vine and some other nocturnal, night blooming plants will open only during the night, whether or not they are placed in a dark closet at the surface of the earth during the daytime, which further suggests that the opening action of these buds is not due to the absence of visible light but possibly to the presence of some form of nighttime radiation.

It therefore becomes apparent that some biological responses in plants react to certain areas of the so-called general nighttime background radiation in a positive way, rather than merely to the absence of the visible light during the dark nighttime period.

A possible explanation for some sort of nighttime radiation that is not present in the daytime might be derived from Van Allen's suggestion that the solar winds, consisting of charged particles emitted continuously from the sun at velocities varying from 670,000 to 1,600,000 m.p.h., compress into a rounded thin layer on the daylight side of the earth and sweep into a long tail on the night side. Van Allen further suggests

127

that the earth's magnetic field causes a positive electrical charge on the morning side of the boundary and a negative charge on the opposite or evening side.

After one of my lectures with the time-lapse pictures, this time to the staff of a research laboratory in the New England area, one of the laboratory technicians told me of an experiment that he was working on, designed to study the normal nighttime activity of certain nocturnal rodents. He mentioned that the experiment was being duplicated by other scientists at a nearby university, and the amount of nocturnal activity in the animals differed significantly between the two locations.

It was thought that the difference in the amount of general background noises at the two locations might account for this difference, and tape recorders had been kept going all night at both locations to record any noises that might disturb the animals. The tapes revealed that both locations were very quiet, and that there was no noticeable difference in the noise level or sounds. I asked about the type of construction of the other building and where the experiment was located within the building. I was particularly interested in learning that the other experiment was in the basement of a four-story concrete and brick structure and that the activity of the nocturnal animals was much less there than with the experiment carried on in the building where I was giving the lecture—a one-story, temporary frame building.

This was at least a suggestion of the possibility that these nocturnal rodents would wake up at night and be more active as the result of some type of radiation capable of penetrating into the so-called "dark" animal rooms, and that more of this radiation might be penetrating the roof and walls of the frame one-story building than to the basement animal rooms of the four-story brick and concrete building.

What is possibly of great significance is the indication of biological responses in both plants and animals to such minor variations of extremely low levels of radiation. Here again, it is customary to think of general background radiation only as natural radiation from the sun and the stars and outer space. In our modern civilization there are increasing amounts of man-made radiations present in far greater intensities than the natural background radiation. A serious question now exists as to whether or not this artificially caused radiation may be causing biological responses not only in plants and laboratory animals, but in man.

However, consideration must also be given to the cumulative effects of so-called "insignificant" amounts of radiation from a variety of different sources. In addition to TV sets, other electrical devices capable of producing various types of radiation include microwave ovens, long distance telephone microwave relay towers, police, weather and airport radar systems, nuclear generating stations, atom bomb tests, medical and dental X-ray machines, fluoroscopes, diathermy, radioactive isotopes, electron microscopes, some types of computers and office machines, high voltage electrostatic air filters and AM-FM radio and TV broadcasting stations.

Too often these low levels of added radiation are shrugged off as amounting to no more than "background" radiation, but when all are added together, they can soon double or triple the so-called "normal back ground" level and, as we have seen, sometimes these added low levels turn out to be not as low as originally thought. There is still much left to be learned about the biological effects of this so-called "background" radiation, and how much it may be increased without also having its adverse effects on man's behavior, physical conditions and thought processes.

Twelve

THE TV RADIATION STORY

The November 6, 1964, issue of *Time* carried a provocative article entitled *Those Tired Children*. It told of the report presented by two Air Force physicians at a meeting of the American Academy of Pediatrics in New York City. No explanation for the symptoms of the thirty children being studied could be found after doing all the usual tests for infectious and childhood diseases. Both food and water supplies were checked. The symptoms included nervousness, continuous fatigue, headache, loss of sleep and vomiting. Only after further checking was it discovered that this group of children were all watching television three to six hours a day during the week and six to ten hours on Saturdays and Sundays.

The doctors prescribed a total abstinence from TV. In twelve cases the parents enforced the rule and the children's symptoms vanished in two or three weeks. In eighteen cases the parents cut the TV time to about two hours a day and the children's symptoms did not go away for five or six weeks. But in eleven cases the parents later relaxed the rules and the children were back again spending their usual time in front of the picture tube. Their symptoms returned as before.

The report concluded that TV watching in itself is not necessarily bad, but that some children become addicted to it and fall into a vicious cycle of viewing for long hours and thus become too tired to do anything else. Other reports have suggested over-psychological stimulation in children from the program content of too many western thrillers and murder mysteries. Little or no consideration seems to have been given to the

130

question of possible radiation exposure. However, epileptic seizures in some children have been reported as being caused by the flicker from TV sets and some further questions have been raised regarding possible effects of sonic energy.

In order to determine if there might be any basic physiological responses in plants or laboratory animals to some sort of radiation or other form of energy being emitted from TV sets, we set up an experiment using a large-screen color TV. One-half of the picture tube was covered with one-sixteenth inch solid lead, customarily used to shield X-rays, and the other half was covered with ordinary heavy black photographic paper that would stop all visible light but allow other radiation to penetrate. Six pots, each containing three bean seeds, were placed directly in front of the portion covered with the photographic paper, six more were placed directly in front of the portion covered with the lead shielding, and another six pots were placed outdoors at a distance of 50 feet from the greenhouse where the TV set was located.

At the end of three weeks, all the young bean plants in the six pots outdoors and the six pots in front of the lead shielding showed approximately six inches of a normal appearing growth. All the bean plants in the six pots shielded only with the black photographic paper showed an excessive vine-type growth ranging up to 31-1/2 inches. Furthermore, the leaves were all approximately 2-1/2 to 3 times the size of those of the outdoor plants and those protected with the lead shielding. The bean plants in front of both the black paper and the lead shielding that were placed at the highest point (so that the bottom of the pot was approximately in line with the top of the TV set), showed considerable root growth emerging from the top surface of the soil. The plants in front of both the black paper and the lead shielding, directly in front of the center horizontal line of the picture tube or near the bottom of the TV set, and those at a distance of 50 feet, showed no such upward directional growth of the roots, causing them to emerge from the top surface of the soil in the pots.

(I later learned, in talks with scientists at the United States Aerospace Medical Center, that wheat seedlings, orbited in a biospace capsule, had behaved in a strikingly similar manner. The random growth of the wheat was thought to be due to weightlessness, but no such condition applied to the bean seedlings. A more logical explanation for the wheat might lie in the fact that the space capsule was being bombarded from all directions by radiation, as were the beans in the pots.)

Such hard-to-explain results prompted setting up a similar experiment using white laboratory rats. Two rats, approximately three months old, were placed in each of two cages, directly in front of the color television tube, and the set was turned on for six hours each weekday and for ten hours on Saturday and Sunday. One cage was placed in front of the half of the tube covered with black photographic paper and the other cage was placed in front of the lead shield, which was, in this instance, increased to 1/8- inch thickness. Lead was also placed under the shielded cage, around both sides, and extended higher in the back between the cage and TV set. This was done in order to assure more complete shielding than that given the bean roots which showed more random directional growth when one flat piece of lead was used to cover half of the picture tube. The sound was turned off but it should be pointed out that turning off the audible sound does not rule out the possibility of sonic energy in the range of 15 kilocycles, which, in some circumstances, can be produced by the action of the picture scanning device.

The rats protected only with the black paper became increasingly hyperactive and aggressive within from three to ten days, and then became progressively lethargic. At 30 days they were extremely lethargic and it was necessary to push them to make them move about the cage. The rats shielded with the lead showed some similar abnormal behavioral patterns, but to a considerably lesser degree, and more time was required before these abnormal behavioral patterns became apparent. This experiment was repeated three times and in each instance the same results were obtained. The lesser degree of response noted in the animals shielded with lead may have been due to another TV set located six feet away which was, at the time, considered to be a "safe" distance. The second set was black and white.

When the first color television set was placed in the greenhouse area of our laboratory, the location was 15 feet from our animal breeding room, with two ordinary building partitions in between. We observed that immediately following the placing of the color television set in the greenhouse, our animal breeding program—which had been going on successfully for over two years—was completely disrupted, and litters of rats which had previously averaged eight to twelve young immediately dropped off to one or two, and many of these did not survive. After the TV set was removed, approximately six months were required for the breeding program to return to normal.

After the second TV set was in operation all the young rats in one of the cages died within ten to twelve days. Two of the rats that appeared extremely lethargic and almost dead were taken to the animal pathology laboratory of the Evanston Hospital where they soon died. Autopsies were immediately performed. Microscope slides were made and the autopsy report indi-

133

cated brain tissue damage in several instances. As I had additional opportunities to speak and show the time-lapse films, I also made a point of showing photographs of the rat brain tissue to several other scientists who were outstanding specialists in this field of brain research. One doctor on the west coast confirmed the original report, but another doctor at a different university disagreed. At two more research laboratories on the east coast there was similar disagreement. Another doctor at a highly regarded medical center said he thought one microscope slide possibly showed brain tissue damage, but an official report from the Radiological Division of the United States Public Health Service gave the opinion that the defects or imperfections noted in the brain tissue slides were artifacts made in the tissues at the time the slides were prepared. This is quite possible and does occur sometimes when microscope slides of such delicate tissues are being made. In the *Psychological Bulletin,* Vol. 63, No. 5, May 1965, an article entitled Behavioral *Biophysics,* by Allan H. Frey, of the Institute For Research at State College in Pennsylvania, states:

> Radiated energy in the electromagnetic (EM) spectrum is an important factor in the biophysical analysis of the properties of living systems. This energy is being used as a tool, both by study of its emission by living organisms (in the micron, millimetric, and centimetric wavelength) and also by applying it to the organism (living organisms absorb, transmit, or reflect it as a function of wavelength). Recent experimentation of these latter wavelengths is becoming of interest to psychologists because of behavioral implications.

Dr. Susan Korbel, at the University of Arkansas, has reported laboratory rats "dancing around" and acting "as though they had been given a type of nerve gas used in World War I" when they were subjected to low levels of microwaves. There have also been reports from Manitoba, Canada, of dairy herds, located within two miles of telephone microwave relay towers, giving considerably less milk, poultry producing only a fraction of their usual egg quota and flocks of chickens going into sudden, unexplained hysterical stampedes.

There have been an increasing number of reports, too, which deal with problems such as the difficulty of maintaining discipline with all ages of school children, their lack of ability to concentrate and forms of lethargy which include a deep, abnormal type of sleep. While it may be argued that modern psychiatric methods have made it possible to detect these problems at earlier stages, we still have had a tremendous increase in the crime rate, violence, and rioting among the youth of this country.

The practice of administering behavioral modification drugs, or "peace pills," as they have sometimes been called, to grade school children has caused much controversy and concern, not only on the part of parents but also among many congressmen, government officials and physicians. This hyperactivity problem may well be the result of exposure to radiation from television sets, to which children are particularly susceptible.

I did manage to show the time-lapse films including the pictures of the bean plants and white rats to the research and engineering people at two of the large television manufacturing companies. This ended all communication with one of the companies and, with regard to the other, I received the following letter from the Electronic Industries Association:

August 6, 1965

Dear Mr. Ott:
  Mr. _____ has been kind enough to give us copies of

135

his correspondence with you in connection with possible radiation hazards from television set viewing.

We can confirm . . . that the problem of radiation protection is one of industry-wide concern. Engineering Committees of the Electronic Industries Association as well as the manufacturers themselves are active in the preparation of standards to insure maximum reproductability and accuracy of ex-radiation measurements. These measurements are made by manufacturers of television sets to insure conformity with the current exposure standards of the National Committee on Radiation Protection and of the International Commission on Radiation Protection. These bodies have set a limit of 0.5 mr/hr measured at 5 centimeters from the surface of the set. At this level, no detectable somatic injuries were expected even if the level were to be exceeded by a factor of 100.* There is good evidence that television sets made in this country meet these current standards.

It is difficult, therefore, to offer any scientific explanation for your reported observations on the basis of exposure to radiation from television sets. For this reason, any evaluation of the significance of your findings will have to await a complete report of your work. Should there be any indication that present-day radiation protection standards are inadequate, you can be assured that we are most anxious to see that remedial action is taken.

> Very truly yours,
> Jack Wayman
> Staff Director
> Consumer Products Division
> Electronic Industries Association
> Washington, D. C. 20036

In order to determine if television sets emitted any harmful X-rays, and for the stated purpose of attempting to duplicate the results of my experiments with the beans and white rats, the Radio Corporation of America

*Indicates 50 mr/hr were considered safe in 1965.

retained the Bio-Analytical Laboratory, Freehold, New Jersey, an independent research laboratory, to test two of their sets. My offer to the director of research for R.C.A. to work with the Bio-Analytical Laboratory and show them exactly what I had done and to test the TV set that I had used was not accepted. Later, the Bio-Analytical Laboratory issued a report stating that no abnormal biological effects were found in either white rats, bean plants or *Tradescantia* placed in front of either of the R.C.A. sets that were tested. Following this report, the director of research for R.C.A. was quoted in the press as saying, "The matter of the possibility of any harmful radiation from TV sets was under complete control by the entire industry and sets were constantly being tested and double checked for any possible X-rays by the Underwriters Laboratories." He was further quoted as saying, "It is utterly impossible for any TV set today to give off any harmful X-rays."

However, considerable confusion apparently existed. When a representative of Underwriters Laboratories was questioned by the press on this point, he said that they had tested for electric shock, etc., but he flatly denied that UL had tested for possible X-rays emitted from the sets.

Soon afterward, the General Electric Company recalled thousands of their color TV sets, announcing that they were defective and did give off some X-rays, but not enough to cause concern, and that the problem was being taken care of by their service men.

Next in the sequence of these events was an announcement by the United States Surgeon General that the radiological division of the United States Public Health Service had measured various models of television sets from a number of manufacturers, and that the matter of X-ray emission seemed to be more of an

137

industry-wide problem than the defective sets of General Electric had indicated originally. Measurements made by the U.S. Public Health Service indicated variations in the amount of X-rays from similar models manufactured by the same company, and it is recorded in the Congressional Record of the Hearings of the Committee on Interstate and Foreign Commerce, House of Representatives (pages 384-385), that the highest level measured in any particular tube was 800,000 milliroentgens per hour, or 1.6 *million* times the acceptable safety level of .5 mrh established by the National Committee on Radiation Protection.

The testimony presented at the Congressional hearings produced much valuable and helpful information, including some fundamental facts that had apparently been overlooked. It was pointed out that sets designed to operate on line voltage of 115 volts might be within the X-ray safety limitation of .5 mrh, but could start producing excessive amounts of X-ray if the line voltage fluctuated above the stipulated voltage of 115 volts. It is not uncommon for the line voltage to fluctuate considerably between the peak load and off peak periods, and a maximum voltage of 130 volts is quite acceptable.

The comparatively recent introduction of such subjects in research as the wavelength resonance of biological oscillators, conductors, and photoreceptor mechanisms not only opens new avenues of approach toward a better understanding and explanation of living matter, but also points up some of the fallacies of the past. For example: it has been common practice in establishing radiation safety levels to refer to the level of natural background radiation for comparison. However, the instruments used for measuring are only capable of recording total energy and do not show any breakdown or distribution as to intensities of specific wavelengths or frequencies. General natural background radiation represents a low level of evenly distributed energy in a broad background energy spectrum. The X-ray radiation from a TV tube is contained in a very narrow spike within the range of less than one angstrom unit.* Therefore, the intensity of the radiation in this narrow band of X-ray would have to be extremely high in order to equal the total energy of the broad, even distribution of total background radiation. Biological systems sensitive to this narrow spike of X-ray radiation from the TV tube would therefore be greatly over-stimulated.

A rough comparison would be the difference in the effects of the amount of sunlight on the total surface of a magnifying glass compared with the same amount of light energy concentrated down to a pinpoint. However, the light concentrated from the magnifying glass is directional, whereas the concentrated spike of X-ray from a TV tube is not, and may travel in all directions. A baby's crib on the other side of the wall from the back of a TV set could be in a very dangerous position as standard building partitions do not prevent transmission of X-rays. The excessive X-rays from the defective G.E. television sets were stated by the company to be generally directed downward. What effect this might have on people on the floor below has not been determined, but it certainly raises questions about the multiple use of TV sets in hospitals, hotels, motels and especially TV show rooms.

It has been general practice to consider only evidence of visible injury or damage to cell tissue in studying the harmful effects of radiation. However, our studies have shown that the pigment granules of the epithelial cells of the retina, which are recognized as

* Source of information: personal interview and correspondence with Dr. John L. Sheldon, Research Manager, Television Products Division, Corning Glass Works.

having no visibility function, are highly stimulated when placed near a TV tube which has been covered with heavy black photographic paper so that no visible light reaches the cells.

If this layer of cells in the retina which have no visibility function is, in fact, the photoreceptor mechanism that stimulates the pineal, pituitary and other areas of the mid-brain region by means of neurochemical channels, then levels of radiation well below those necessary to produce detectable physical injury to cell tissue could reasonably be expected to influence the endocrine system and produce both abnormal physical and mental responses over an extended period of time. Radiation stress must be considered as a possible variable or contributing factor. Just how the mechanism works that causes certain pigments of some plants, animals and people to react to specific wavelengths within the total electromagnetic spectrum is a challenge to future research.

The indication that radiation stress does occur in biological systems raises serious doubt as to whether the present safety factor of .5 mrh is low enough. With each of the many cuts in the recommended safety level since X-rays were first discovered it was thought that surely the lower figure would be safe. Unfortunately, experience has shown otherwise, and what additional knowledge has been gained has been learned the hard way as the result of X-ray injury to many people.

Photobiological responses through the extremely sensitive retinal-hypothalamic-endocrine system being directly exposed for extended hours of television viewing, especially at close range, to even trace amounts of radiation, may be compared to trace amounts of one part in ten million in chemistry, and may point to safe levels of radiation as being in a similar magnitude of one ten millionth mrh, or less.

On April 24, 1970, an excellent article by staff writer Ben Funk was released by the Associated Press. The article presented a review of the TV radiation problem, and is reprinted here with permission.

The Battle Against TV Radiation—It's Just Begun
by Ben Funk

WASHINGTON—Three years after the first disclosure that some color television sets were bombarding viewers with X-ray beams, science has begun to define the resulting dangers. But it may be years before radiation is banished from living rooms.

The news in 1967 alarmed many TV watchers, sparked a Congressional investigation, and led to the passage of the 1968 Radiation Control Act, setting limits on rays receivers may emit.

But millions of sets are not covered by the law and new discoveries indicate that government standards are too low to assure full protection.

The standards, to be fully applied June 1, 1971, require that no TV set may spill out more than 0.5 milliroentgens of radiation per hour—a level considered safe at the time the standards were drafted.

However, recent findings by scientists in the Department of Health, Education and Welfare (HEW) indicate that X-ray emissions below the 0.5 level and on down to zero penetrate body tissues with subtle but harmful effect.

The only answer to the problem, says Dr. Arthur Lazett, assistant director of HEW's Bureau of Radiological Health (BRH) is to "eliminate radiation entirely" from the receivers.

Even if this ideal were realized, the problem would linger. Of 30 million color TV sets sold through 1969, about 25 million are still operating and many will last for years. The Electronic Industries Association, which represents most manufacturers, says about 6 million sets will be sold this year.

"The standards are too low," consumer advocate Ralph Nader told Chairman Warren Magnuson, D-

141

Wash., of the Senate Commerce Committee, which recently held hearings on the act. "Millions of people are being exposed to the risk of physical, genetic and eye damage."

Until recently, official ignorance about TV radiation dangers hampered the attack on the problem.

For example, when several surveys turned up sets spewing heavy rays, the U.S. Public Health Service warned viewers on April 16, 1969, to sit no closer than six to ten feet from the sets. But it could not say why this was judged to be a safe distance, nor what harm the viewer could expect if he sat nearer.

Five months later, the Federal Trade Commission issued an identical warning. Asked why the FTC also failed to spell out the consequences, a spokesman said science had not calculated the injurious effects of prolonged exposure to TV rays.

Since then, new light has been thrown on the subject by a scientific team of the Bureau of Radiological Health.

When the research began, Dr. H.D. Youmans said the problem was approached with some skepticism. "We questioned whether TV radiation was important, because it was so low compared to the output of an X-ray machine," Youmans said. "We thought the rays would be soft and nonpenetrating.

"Instead, we found rays escaping from the vacuum tubes to be harder and of higher average energy than we expected. They penetrated the first few inches of the body as deeply as 100-kilovolt diagnostic X-rays. You get a uniform dose to the eye, testes and bone marrow."

Dr. Norman Telles said the team also speculated at first that there was a threshold below which radiation ceased to penetrate. Now, he reports, "we have made the assumption that there is no threshold, that radiation down to the zero level evokes a response from body tissues."

At a congressional hearing a year ago, Dr. Robert Elder, director of the BRH, testified that small doses of radiation are cumulative and may cause genetic damage affecting future generations.

Rep. Paul Rogers, D-Fla., co-author of the Radiation Control Act, says a Sarasota naturalist, John Nash Ott,

142

"got us started in 1967" on the road toward control of radiation from electronic products.

Ott reported at the time that a cage of young rats placed close to a color TV set, with the sound off and the picture tube covered with black photographic paper, became highly stimulated, then progressively lethargic, and all died, in ten to twelve days.

"We brought him to Washington for a briefing," Rogers said, "but we didn't say anything for a while. We were afraid we would scare a lot of people. And when we checked the U.S. Surgeon General, he told us there was nothing to worry about."

General Electric Co. announced May 18, 1967, that it was recalling 154,000 sets giving off excessive radiation. The shunt regulator tubes, which control the high voltage centering the electronic beam to the picture tube, were poorly shielded and located. A few sets were delivering as much as 40,000 milliroentgens.

As Congress and public health offices around the nation were pelted with inquiries from concerned set owners, the Electronics Industries Association declared that the GE problem was "an isolated case."

But the following December, a U.S. Public Health survey of 110 sets in St. Petersburg, Florida, found that 18 were emitting rays in excess of safety standards. They included eight makes.

With this evidence, Rogers' House Subcommittee on Public Health and Welfare moved into hearings that were to lead Congress into the whole field of radiation from electronic products.

"Color TV was the glamor symbol," Rogers said. "There are so many of them."

As the hearings opened, the National Center for Radiological Health checked 1,124 color sets in Washington D.C., and found 66 of various brand names putting out excess radiation. The amounts ranged up to 25 times the 0.5 level.

Then the Suffolk County Public Health Service of Long Island, New York, reported in April, 1969, that 20 percent of 5,000 sets examined over a two-year period were emitting dangerous rays.

143

Color receivers require in the neighborhood of 25,000 volts. When voltage goes beyond this level radiation builds up rapidly and escapes from all sides of the set and the bottom if shielding is insufficient to contain it.

It was found in the surveys that repairmen often boost the voltage to brighten and sharpen the picture and sometimes damage or fail to replace the shielding. Radiation which ranged up to 150 milliroentgens from some Suffolk County sets fell sharply when repairmen called out by the county lowered the voltage.

The Radiation Control Act passed by the House 381 to nothing, was cleared by a voice vote in the Senate after only a brief discussion, and was signed by former President Johnson in October, 1968. It authorized HEW to name a 15-man committee, equally divided among the industry, the public and the government, to draft standards.

Government scientists proposed an 0.1 milliroentgen limit on radiation, but the committee rejected the limit as too tough on the industry and settled for 0.5.

The standards were set up in three stages. The first required only that new sets meet the 0.5 limit. The second indicated that receivers manufactured after June 1, 1970, must remain within this limit even when controls are maladjusted in a way that would increase the rays.

As of June 1, 1971, the standard will be broadened to cover not only maladjustment of controls, but also component or circuit breakdowns.

Nader charged earlier this year that the Radiation Control Act is not working because "the forces of industry and bureaucracy have prevailed."

J. Edward Day, special counsel for the Electronics Industries Association, told Representative Rogers' committee that TV makers are moving in a variety of ways toward cleaner sets.

"These efforts are being pursued," he said, "not because there is any feeling on the part of TV manufacturers that a hazard situation exists or that there is any justifiable cause for public alarm." It is an effort, he said, "to bring an end once and for all to the flurries of public excitement over TV radiation."

It was a great relief to have some official confirmation of possible X-ray hazards from TV sets. Friends and relatives were polite and understanding but possibly a bit fed up with all my warnings about radiation from TV. It was hardest for the grandchildren to understand why only the mice should be allowed to watch the new color set. One grandson who lived in Rome, Italy, after a visit to Grandma and Grandpa, was asked by his teacher at school to tell about his trip to America and what his Grandpa did. The next day the teacher called his mother to mention what a fanciful imagination the child had; but did his Grandpa really raise mice and keep the only color TV set out in the mouse house for them to watch?

The first public showing of the pictures of the responses of the bean plants and white rats in front of the TV sets was on October 3, 1966, to a meeting of the 100th Technical Conference of the Society of Motion Picture and Television Engineers in Los Angeles, California. The pictures fitted in well with the pumpkins and other segments, but parents especially seemed more concerned with the idea of their children sitting close to the set and staring directly into the picture tube, (which like an X-ray machine, consists of a cathode ray gun operating on approximately 27,000 volts of electricity) and shooting a stream of electrons directly at their eyes. When these electrons hit a metal target they produce X-rays.

The problem is, therefore, to shield the viewer from these electrons and X-rays so that it is still possible to see the picture. X-rays are able to penetrate not only human flesh, but also can penetrate steel and still affect the chemicals on a photographic plate. This is how X-ray pictures are made. A question facing us today is whether all the chemical effects of X-rays on human tissues are known.

What really concerns me in the above context is the effect of the very low levels of radiation that can influence the pattern of the streaming of the chloroplasts in the cells of a leaf, or the pigment granules in the epithelial cells in the retina of the eye, without showing any evidence of cell injury or damage. The roots of the bean plants that grew upward looked perfectly normal and so did the white rats that were dancing around and attacking each other. Is it possible that these very low levels of radiation affect the behavioral patterns and learning abilities of children without producing any signs of physical injury or cell structure damage?

The main difference between black and white and color TV is that the black and white sets have only one cathode gun, whereas a color set has not only three cathode guns, one for each of the primary colors, but also generally operates in a higher voltage range than black and white sets. I am constantly asked what is a safe distance for children to sit when watching TV, but when I think of the rat breeding colony being so completely disrupted at a distance of 15 feet with two intervening building partitions, I can only answer that I really don't know, but that the distance that might be considered safe would undoubtedly vary with different sets.

Thirteen

TRACE AMOUNTS OF RADIATION AND FULL-SPECTRUM LIGHTING

My wife and I had been spending more and more time each year in Florida and we finally decided to make Sarasota our permanent home. Just before signing the contract for the sale of our home in Lake Bluff, Illinois, I had a call from Howard Koch of Paramount Pictures Corporation in Hollywood. They were ready to start work on a new film to be called *On A Clear Day You Can See Forever*, starring Barbra Streisand. The story was based on the Broadway production of the same title, and begins with Miss Streisand singing as she plants various flowers in pots in her garden on the roof of her apartment building, where she also has a small, lean-to greenhouse.

According to the story, Miss Streisand had unusual powers, including making flowers actually grow as she sang to them. The motion picture allowed for the use of time-lapse photography to show the flowers growing, which of course could not be done on the Broadway stage. Mr. Koch asked me if I could make the necessary time-lapse films of the flowers, which would include geraniums, roses, iris, hyacinths, tulips, daffodils, and possibly other varieties. Moreover, I would have to start at once if it was to be done at all that year because of the seasonal requirements for transplanting some of the plants, and especially the chilling or cooling requirements of all the spring bulbs during the normal winter months.

Such an important project could not have come at a more inopportune moment; to take satisfactory time-lapse pictures of the growth of such flowering plants from the time of emergence of the first shoots from the ground required the use of the ultraviolet transmitting plastic greenhouse and I had already decided that mine was due for replacement.

The mechanism of the large shutters that closed each time a single exposure was made in any camera was worn out, and the plastic itself needed replacing after approximately twenty years exposure to the weather. Even if I should try to move the old greenhouse to Florida, it would take months to get it back into working order, and there simply wasn't time to do this and prepare all the plants to be photographed at the same time. I therefore put a clause in the sales contract for our residence granting me occupancy and use of the old greenhouse for one additional year if necessary, to do the work for the film.

Meanwhile, I had assisted one of the light bulb manufacturers in developing a new, full-spectrum fluorescent tube with added ultraviolet, to duplicate as nearly as possible the natural spectrum of outdoor sunlight. The problem of light intensities, however, had to be considered. Full, direct natural sunlight ranges as high as 10,000 foot candles, while the maximum intensity from a fluorescent fixture containing ten 8-foot tubes is only approximately 1,000 foot candles at a distance of 10 to 12 inches. I immediately set up an experiment in a windowless room in my Florida office space, to see whether such sun-loving plants as roses and geraniums would grow under as little as 1,000 foot candles of fluorescent light even if it had the ultraviolet wavelengths added—those not normally present in artificial light sources.

While this experiment was in progress, I began building work tables and installing some of the time-lapse equipment brought down from Lake Bluff. Only if the flowers would grow under the new full-spectrum fluorescent lights could I possibly complete the project in Florida in time to meet the deadline. There was not even sufficient time to build a make-shift greenhouse, with the necessary shutters that must close each time a picture is taken in order to shut out the natural sunlight and obtain an even exposure using the same artificial sources, day and night, on dark days and bright days. But I could very quickly set up some automatic timers and the new lights, which I suspended to cables fastened to a winch. In this way I could keep the lights as low as possible over the growing plants and raise them as the plants grew taller.

The pilot experiment to see if roses and geraniums could be grown successfully under only 1,000 foot candles of the new type full-spectrum lights was entirely successful. I was ready to go full steam ahead with the whole project. First, I bought some old refrigerators, so that I could plant all the spring bulbs in pots and put them through the equivalent chilling of a normal winter up north. This meant leaving the old family car out in the salty air and giving the garage to the refrigerators full of bulbs.

To try to cover all contingencies that might arise to interrupt the shooting of any particular sequence, and to have a selection of several "takes" of each particular flower, it is necessary to make several complete sequences of each subject and have additional subjects timed to come into bloom at regular intervals, so that if a hurricane, a power failure or any other catastrophe spoiled one picture at a crucial moment, there would always be a "stand-in" plant. As a further precaution, in case my refrigerators failed to duplicate in Florida the chilling effects of a northern winter, I also made arrangements with one of the northern bulb growers to set aside a number of pots of all of the varieties of bulbs we were using, so that they could be shipped to Florida

149

at the last minute if necessary.

Everything proceeded smoothly and all the flowers grew well, but I noticed that the results were better with plants placed near the center of the fluorescent tubes than with those placed under the ends of the tubes. This could have been due to the higher intensity of light that the plants received near the center, although I had white reflectors on the walls at each end of the light fixtures to make up as much as possible for the loss of intensity—otherwise the plants near the ends of the tube would get the light only from directly overhead and from one side, instead of from both sides as did those in the center.

However, in view of the work I had done in studying the biological responses of plants to trace amounts of radiation, I was suspicious of the concentration at the ends of the tubes. Each fluorescent tube works on the same general principle as the cathode guns in a television tube or the tube in an X-ray machine. The principal difference is that they operate at much lower voltages, and according to all the textbooks on the subject, at voltages low enough so that it is impossible for them to produce any harmful type of radiation. Nevertheless, the plants growing under the ends of the fluorescent tubes did not develop as well as those under the centers, so I set up an experiment with beans to determine whether there might be trace amounts of radiation from each cathode that would cause abnormal biological responses in plants.

The results were quite startling. The bean plants growing close to the ends of the 8-foot tubes were noticeably stunted in their growth, whereas those near the center of the tubes, or at a distance of ten feet away from the tubes, appeared quite normal. Some pots containing the seeds were placed very close to the ends of the tubes, others close to the centers. I checked temperatures with a thermometer and found that there was actually more heat coming from the ballasts in the fixtures near the centers of the tubes than from the cathode ends, indicating that the stunted growth of the seedlings near the cathodes was not caused by excessive heat. I then repeated the experiment with two pots, shielding one pot from the tubes with a 1/16 - inch lead sheet, and only black paper between as a shield for the other. Both pots were placed side by side where the ends of both sets of fluorescent tubes met. The beans in the pot shielded only with paper showed the same stunted type of growth noted in the first experiment, whereas the seeds in the lead-shielded pot grew normally.

I then decided to set up still another experiment in the windowless time-lapse room, using the same two fixtures, each holding ten 8-foot tubes and again placed end to end. Bean seeds were placed on wet cotton pads in plastic dishes which were then placed in similar locations in close proximity to the ends of the two banks of fluorescent tubes. The results of this experiment showed that the bean seeds would sprout, but the little roots would immediately start off in a more random directional growth, with many of them pointing upwards. However, when a thin lead shield was placed around two inches of each end of the fluorescent tubes so as to cover the cathodes and thus shield the bean seeds, the roots would grow downward in the normal way.

Immediately after the bean seed experiments, one of the two identical, large fluorescent fixtures with its ten 8-foot tubes was taken to a laboratory specializing in testing electronic equipment for radiation emissions. Their report indicated that neither the fluorescent fixtures nor the tubes generated any measurable ionizing radiation, electromagnetic radiations between 1

151

and 12,400 megahertz, or ultraviolet radiation in the frequency range high enough to cause air ionization. These are the areas of the electromagnetic spectrum customarily checked by various governmental agencies when testing for radiation emissions from television sets, microwave ovens and other household appliances, but these tests do not rule out the possibility of radiation emissions in other parts of the total electromagnetic spectrum.

This study would seem to indicate that the bean seeds and plants may be more sensitive to trace amounts of radiation than standard present day radiation measuring equipment. This may be due to the cumulative effect of radiation on biological systems, whereas the meter readings of electronic instruments in common use today read only a given level of radiation at any particular moment.

Perhaps bean seedlings will have to be used for detecting trace amounts of radiation until more sensitive measuring devices are developed, just as canary birds were used for many years in coal mines to give warning of the presence of coal gas.

Fourteen

PHOTOBIOLOGY COMES OF AGE

After our move to Florida in 1966, I felt the need for a more substantial structure from which I could disseminate research findings, and so we decided to establish the "Environmental Health and Light Research Institute." A group of interested citizens of Sarasota agreed to serve on the Board of Trustees, and a number of prominent physicians are serving on the Medical and Scientific Advisory Board. The Environmental Health and Light Research Institute became associated with New College of Sarasota, and it is planned ultimately to build a permanent light research facility.

In addition to my work with the Duro-Test Corporation in the development of a new, full-spectrum fluorescent light, I think I can claim some credit for persuading several of the large plastic manufacturing companies to start making ultraviolet transmitting plastic material for windows and skylights, and, especially, for making full-spectrum spectacle lenses and contact lenses available to the public. Both spectacle and contact lenses are now available in clear and neutral gray. When sunglasses are needed, these neutral gray full-spectrum lenses will reduce the light intensity proportionately through the ultraviolet, visible, and infrared wavelengths, whereas many other types of sunglasses are designed specifically to cut out all of the ultraviolet and infrared, and then, depending on their color, give a considerably distorted spectrum of wavelengths.

Much of the commercial gray glass available, especially for large tinted windows, does not cut down the intensity of the wavelengths in the visible spectrum evenly, but principally lets through three bands of wavelengths which, when combined, look gray to the eye, but do not give the more continuous even spectrum of the neutral gray full-spectrum plastic. For this reason, some of the plastic lens companies are using our Full Spectrum Symbol of Approval, which sets forth rigid standards which must be met before it can be used on any product. This is designed to protect the public from inferior products, and companies using this symbol pay a small royalty which helps support further light research.

With the greatly increasing interest in environmental problems, these new full-spectrum products are coming into general usage. The increased interest in ecology has dealt primarily with man's environment and the problems of air and water pollution. But to these must be added the problem of polluted light, as well. Air pollution not only causes respiratory problems but also further pollutes our light environment. Scientists at the Smithsonian Institution in Washington, D.C., report a loss there of 14 per cent in the over-all intensity of sunlight during the last sixty years. Scientists at the observatory on Mount Wilson in California report not only a loss of 10 per cent during the last fifty years in the average intensity of sunlight even at that high elevation, but a 26 per cent reduction in the ultraviolet part of the spectrum. In New Jersey, farmers have been reporting difficulty in growing squash because of increasing virus problems, which are thought to be spread by insects, especially aphids that infest the squash plants. This problem seems to have been increasing in spite of higher applications of pesticides. Scientists from both the United States Department of Agriculture and Rutgers University have now reported the elimination of both the viruses and the insects and a fivefold increase in the yield from squash plants as the result of spreading aluminum foil on the ground underneath the plants. Aluminum foil is a very good reflector of visible light and a particularly good reflector of ultraviolet wavelengths.

This evidently beneficial effect from intensifying sunlight, and especially the ultraviolet wavelengths, to make up for what is lost from the filtering effect of air pollution, brings to mind another somewhat similar situation.

While the project was under way at the Northwood Mink Farms, Mr. Grosse one day pointed out quite a large section of one of the sheds where, for the past seven years, their records indicated that the average number of young was greater in each litter and that the pelts consistently graded higher than normal. No explanation for the better results could be found, but if all the animals could be made to respond in the same way it would mean a substantial increase in profits. The food, water and cages were exactly the same for all the animals. The construction of all the sheds was exactly the same, but when a number of females with a record of the smallest litters and poorest quality fur were selected from other sheds and placed in this particular area, they too showed a significant improvement in the condition of their fur as well as an increase in the average number of young in ensuing litters. This truly baffled Grosse and his staff.

In looking over the situation, I happened to notice a building adjacent to this special area that had aluminum siding, whereas all the other adjacent buildings had corrugated iron siding. Corrugated iron is a poor reflector of ultraviolet and aluminum is one of the best. I made a special point of going back the next day with an ultraviolet light meter and found decidedly higher levels of ultraviolet being reflected from the aluminum siding of the adjacent building into the area where the mink had been doing so exceptionally well.

155

If air pollution continues to get worse, perhaps we may soon be wearing aluminum collars to reflect the diminishing sunlight. The loss of 10 to 14 per cent of visible sunlight and even more of ultraviolet is frightening, but civilized man has cut himself off from much greater percentages of sunlight by living indoors behind walls and glass. He has developed artificial sources of illumination that are gross distorters of the visible light spectrum of natural sunlight and almost totally void of any ultraviolet. The number of people wearing eyeglasses and contact lenses is steadily increasing so that even when they are out of doors, most of the ultraviolet is blocked from entering their eyes. Tinted lenses and dark colored sunglasses further grossly pollute the light. The question then arises as to what extent this artificial or polluted light environment may be affecting man's general health and well-being.

In April of 1970, the Kline Chinchilla Research Foundation at Utica, Illinois, announced the results of a five-year study in which the Environmental Health and Light Research Institute and more than 2,000 chinchilla ranchers throughout the world participated. The final results indicated that when ordinary incandescent light was used in the breeding rooms, the litters would average 60 to 75 per cent males and, when "daylight" incandescent bulbs were used, the ratio of males to females would be reversed and average 60 to 75 per cent females.

The Kline Chinchilla Research experiments also indicated that the periodicity of light, and not the temperature, influenced the development of heavy winter fur on the chinchillas. By lengthening the hours of darkness, the animals can be brought into their prime pelt season during any month of the year instead of the normal winter season. The chinchilla industry is now applying these principles on a commercial basis, just as the poultry industry has done for many years in lengthening the daylight hours with artificial lights during the winter to increase egg production.

The results of another interesting study, also published in 1970, indicated that the number and intensity of dental caries in the teeth of hamsters may be linked to the absence of natural or simulated sunlight, according to Drs. R. P. Feller and Spencer W. Burney of the Veterans Administration in Boston, and I. M. Sharon of the School of Dentistry of the University of the Pacific. Their study involved feeding thirty male golden hamsters a high carbohydrate diet containing 60 per cent sucrose. Half of the group was exposed for 12 hours daily to fluorescent tubes with added ultraviolet to approximate natural sunlight, while the other half was exposed to standard cool white fluorescent light. The group under the full-spectrum light averaged 2.2 teeth with cavities, while those exposed to the cool white averaged 10.9 teeth with cavities—which were also ten times as severe. The artificial light which simulated natural sunlight also had an effect on the sexual maturity of the hamsters, according to the report. The development of the male sex organs was only one-fifth as great in those hamsters under the cool white light source as those under the full-spectrum fluorescent tubes.

Another most important paper appeared in *Experientia 26, 267* (1970) entitled *Electrophysiological Evidence for the Action of Light on the Pineal Gland in the Rat,* by A. Newman Taylor and R. W. Wilson of the Department of Anatomy and Brain Research Institute of the University of California School of Medicine in Los Angeles. Doctors Taylor and Wilson implanted a bipolar stainless steel electrode into the rat pineal gland and were able to demonstrate by electrical recordings, from the rat pineal, a high tonic level of spontaneous activity in darkness which is inhibited by periods of illumination, thus confirming the anatomical and functional evidence for the action of light on the pineal. The inhibitory effect of light on pineal electrical activity was found to be mediated by the retina. This evidence of measurable electrical activity in the pineal gland of the rat responding to photic stimuli of the retina further confirms the work of other scientists who have demonstrated the effect of light through the retina by chemical measurement of the production of various hormones by the pineal gland of the rat.

At the American Association for the Advancement of Science meeting in Chicago, December 26-31, 1970, Lewis W. Mayron Ph.D., presented a paper entitled *Environmental Pollution, Its Biological Effects and Impact on the Bioanalytical Laboratory*, in which he explored the biological effects of radiation from television sets and fluorescent lighting tubes. In discussing the published results of our experiments with the bean plants and white rats placed in front of a television set, Dr. Mayron comments, as follows:

> Thus it appears that the radiation emitted from the TV set has a physiological effect both on plants and animals and it is likely that this effect, or these effects, are chemically mediated. If *Those Tired Children* are any indication of a trend, the bioanalytical laboratory may be called upon to chemically determine low-grade radiation toxicity.
>
> Although there is as yet no indication of the body chemicals involved in the physiological effects of TV radiation, there is some indication of a chemical effect of the radiation of Ultra High Frequency (UHF) radio fields. Gordon (*Science* 133: 444, 1961) has found that UHF fields result in the accumulation of acetylcholine along nerve fibers. Korbel and Thompson (*Psychological Reports* 17: 595-602, 1965) reported on the behavioral effects of stimulation by UHF radio fields, which just happens to

correlate with the behavior of the rats in front of the color TV screen and which also correlates with behavioral effects due to the accumulation of acetylcholine. Acetylcholine in small concentrations leads to cholinergic hyperactivity, while larger concentrations lead to a decrease in activity (Crossman and Mitchell, *Nature* 175:121-122, 1955; Koshtoiants and Kokina, *Psychological Abstract* 32:3584, 1957; Russell, *Bulletin British Psych. Society* 23:6, 1954 [abstract]). Nikogosyan found significant reductions in blood cholinesterase activity in rabbits after a program of UHF exposure (in Letavet and Gordon, Eds, *Biological Action of Ultra-high Frequencies,* OTS 62-19175, Moscow; Academy of Medical Sciences USSR, 1960). Thus, perhaps cholinesterase activities ought to be determined on man and laboratory animals exposed to TV radiation. An interesting addendum to this story on TV radiation is a series of experiments performed by Dr. Ott, using bean seeds and seedlings and fluorescent light tubes.

Dr. Mayron then relates the procedures and results of the bean seed experiments with the fluorescent tubes and further comments: "The implications of this are enormous when one considers the magnitude of the use of fluorescent lighting in stores, offices, factories, schools and homes."

Another new and interesting application of light therapy for the treatment of cold sores and fever blisters was developed by a group of scientists at Baylor College of Medicine, and reported by Dr. Troy D. Felber at the 120th annual meeting of the American Medical Association during June, 1971, in Atlantic City, New Jersey.

The new technique consists of applying a certain type of dye to the skin lesion and then irradiating it with an ordinary daylight type fluorescent light. The virus then somehow becomes inactivated through a process described as photodynamic inactivation.

The work has greater implications than for skin infections only because one type of the same virus *herpes simplex*—is believed by many scientists to cause cancer of the cervix.

A more definite indication of a direct carcinogenic relationship between light and the effects of certain chemicals was reported in Japan. (*J. Genetics,* Vol. 44, 231-240: 1969, S. Takayama, Y. Ojima, Biol. Lab., Fac. Sci., Kwansei Gakuin Univ.; Nishinomiya, Hyogo). Cultured cells were exposed to each of 8 polycyclic hydrocarbons, and then illuminated with white light from a tungsten lamp. Five carcinogenic hydrocarbons were found to be much stronger in photosensitizing activity than three noncarcinogenic ones. Among the former compounds benzopyrene is the strongest and benzanthracene the weakest. As far as the present experimental system is concerned, a clear positive association between photodynamic activity and carcinogenicity was found.

Another interesting study conducted at the Chelsea, Mass., Soldiers Home has shown that the body absorbs calcium more efficiently when individuals are exposed to an artificial sunlight environment as compared to cool white fluorescent tubes. Mr. Luke Thorington, an engineer with the Duro-Test Corp., told of the study at the 1971 Illuminating Engineers Society convention in Chicago.

One of the more interesting developments in medicine during the past few years has been the introduction of light therapy in place of a complete blood transfer for the treatment of *neonatal hyperbilirubinemia,* or jaundice, in premature babies. The use of light for the treatment of jaundice may have been practiced originally in India by mid-wives who placed unclothed jaundiced infants in the sunlight to cure them. The use of artificial light to treat jaundice traces back to the work done in 1958 by Dr. Richard J. Cremer, of Harefield

Hospital in Middlesex, England. Dr. Cremer showed that the *serum bilirubin* levels that cause jaundice could be lowered in infants by exposing them either to sunlight or to artificial blue light. The accepted alternative treatment for severe cases is to perform a complete blood transfusion, which in itself carries considerable risk. In this country, Dr. Jerold F. Lucey, Professor of Pediatrics at the University of Vermont College of Medicine and a past president of the American Academy of Pediatrics, has carried on extensive work in further studying light therapy for the treatment of jaundice in premature babies. Dr. Lucey and his co-workers have experimented with different types of fluorescent lights and recently he has reported to me that he is now using the new full-spectrum fluorescent tubes with added ultraviolet to duplicate the spectrum of natural sunlight more closely.

Several companies are now manufacturing special plastic light therapy isolets, fully equipped with an automatic time clock for controlling the periodicity of the all important light and dark cycle as well as thermostatic control to provide the most desirable temperature for the premature infant. The normal light intensity is usually about 300 foot candles, but may also be controlled as prescribed by the doctor. Three hundred foot candles is roughly the equivalent of deep shade under a big tree outdoors on a sunny day. Full sunlight on a clear day without air pollution ranges around 10,000 foot candles.

When the premature babies are exposed to the light treatment, their eyes are covered with a blindfold as a precautionary measure when the skin is exposed to the light. The treatment is usually for eight hours a day for five or six days. Thus, the benefits of this light therapy for infant jaundice is another benefit mediated through the skin.

Many doctors and hospitals from New York to Los Angeles and around the world have reported enthusiastically on the results of this new light therapy. In a paper presented at a Symposium on Bilirubin Metabolism at the 1970 National Foundation meeting in Chicago, Dr. Jerold Lucey stated that published human clinical experience for several countries with phototherapy now numbers over 5000 cases and no serious toxic effects have been reported to date, even by those most critical and fearful of possible unknown harmful reactions.

Yet fear of possible harmful effects of exposure to ultraviolet (apparently including normal visible light), seems to be of a major concern.

For example, in the August 1970 issue of the *Journal of Pediatrics 77,* No. 2, pp. 221-227, appears an article entitled *"Retinal Changes Produced by Phototherapy."* The authors are Thomas R. C. Sisson, M.D., Stanley C. Glauser, M.D., Ph.D., Elinor M. Glauser, M.D., William Tasman, M.D., and Toichiro Kuwabara, M.D., from the Departments of Pediatrics, Pharmacology and Ophthalmology, Temple University School of Medicine, and Howe Laboratory of Ophthalmology, Harvard University Medical School, Massachusetts Eye and Ear Infirmary. The study was supported in part by three separate grants from the National Institutes of Health, Bethesda, Maryland.

In order to determine if retinal damage occurs during phototherapy of the newborn infant, twelve newborn piglet littermates were continuously exposed for 72 hours to a bank of 10 high intensity blue 20-watt fluorescent lights at a distance of 46 cm (18 inches). The light intensity at this distance was 300 foot candles. A Plexiglas shield .5 inch thick was used to filter all wavelengths below 390 mm; that is all or any ultraviolet that might be emitted from the blue fluorescent tubes.

The right eye of each piglet was dilated daily with atropine, 0.5 per cent, and the left eye was covered with a patch so that 99.5 per cent or better of the light was blocked. Histological observation revealed retinal damage in all the right eyes dilated with atropine and exposed to intense blue light. One piglet lost its eye patch for a period of less than twelve hours during the second day of the experiment and was later found to be clinically blind in both eyes. Except for the piglet that lost its eye patch, no retinal damage was reported in the protected left eyes of the other five experimental piglets. The control piglets were kept in the usual low level illumination of the animal colony and with a diurnal rhythm of 8 hours of light and 16 hours of darkness. Histological observations revealed well developed retinas in all the control animals.

The results of the above experiment were reported under the heading *Phototherapy Exposure Tied to Retinal Damage* in *Pediatric Currents,* Vol. 20, No. 1, Jan. 1971, and included excerpts from several other publications, including the leading article of the *British Medical Journal,* 2;5 1970. A number of possible dangers and other deleterious effects that light might have on the newborn were suggested, although it was further stated that conclusive evidence of these dangers is not available. Of particular concern is the possibility that light may have adverse effects unconnected with bilirubin metabolism such as pineal function, sexual maturation, circadian rhythms and corneal ulceration. It is concluded that neonatal units are probably not justified in adding special lighting apparatus to the already long lists of important equipment needed from limited funds. How curious it is that such articles warn of the many dangers involved with light, but recommend doing nothing about the problem.

Present lighting conditions in the nurseries of hospitals vary considerably as to both intensity and type of lights. Some have windows that permit intensities of daylight that are far greater than the 300 f.c. used in bilirubin isolets. In many baby nurseries the lights are left on all night for the convenience of the nurses caring for the infants, or turned on every time a nurse enters the room. Some nurseries are equipped with short wavelength ultraviolet germicidal lamps to kill airborne germs. Shields are provided as standard equipment in such cases so that no direct ultraviolet light shines on the infants, but I have personally seen installations where the eyes of the infants are exposed to considerable amounts of reflected ultraviolet light, especially from the upper walls and ceiling of the room.

The July, 1971, issue of *Annals of Ophthalmology* carried an editorial, *Light: A Double-Edged Sword,* which noted the work of Noell and Albrecht at the Neurosensory Laboratory of the State University of New York at Buffalo. Quoting from the April 2, 1971, issue of *Science,* it states that:

> Exposure to normal light may result in deterioration of the visual cells and degenerative changes in the underlying pigmented epithelium...in from 7 to 10 days exposure to continuous light of 110 lux intensity from ordinary light bulbs.

However, a study of the complete article reveals additional data. The authors indicate that if a green filter is used over a light source of 1500 lux, severe retinal damage results in 40 hours. Noell and Albrecht further report on the effects of vitamin A deficiency and other chemical reactions within the eye, but conclude that the normal diurnal cycle of light and dark seems to be the essential factor in controlling visual cell viability and susceptibility.

The results obtained by Noell and Albrecht seem to tie in closely with the problems encountered in using a green filter in the light source of the phase contrast microscope when taking time-lapse pictures continuously, day and night, of the pigment granules in the epithelial cells of the retina of a rabbit's eye, and also the chloroplasts in the cells of a leaf of a plant. This further points up the importance of the periodicity of light and darkness on animals as well as plants, and also again reminds me of the problem first encountered with the rosebud refusing to open continuously when the lights were left on all night. Noell and Albrecht report similar retinal damage from exposure to continuous green light not only in several types of rats, but also mice, hamsters and the Galago monkey.

Retinal damage from continuous exposure to light through a green filter in so many different species of animals suggests to me the need for further studies of the effects of placing green or other colored filters in the form of sunglasses or tinted contact lenses in front of the eyes of the human animal. Some further thought might also be given to whether or not such conditions of prolonged exposure to either incandescent or green light can be considered "normal light," especially when such exposure may cause serious damage to the retina. There is certainly a need for further studies of the effect of the periodicity of light and darkness on the eyes of the human animal. Continuous light and the state of being "awake" for seven to ten days would certainly have a deleterious effect on the entire body, not simply on certain cells in the retina. There is still much to be learned about the importance of sleep and its rejuvenating effect on all the individual cells of the entire body.

At the end of the editorial it is stated that the information now available on the indicated harmful effects of ordinary visible light is already being put to use therapeutically in cases of hereditary *retinitis pigmentosa.* By completely excluding light with an opaque flush-fitting scleral contact lens in one eye (in order to preserve one retina), it is hoped to double the patient's visual lifetime. However, before starting any such "protective" therapy, I suggest reading the paper by Chow, K. L., Riesen, A. H., and Newell, F. W., 1957—*Degeneration of Retinal Ganglion Cells in Infant Chimpanzees Reared in Darkness,* in the *J. Comp. Neurol.,* 107: 27-42.

The results of another highly significant experiment that came about somewhat accidentally were reported in the July 30, 1971 issue of *Science.* Dr. Irving Geller, Chairman of the Department of Experimental Pharmacology at the Southwest Foundation for Research and Education in San Antonio, Texas, was subjecting laboratory rats to various types of stress in an attempt to induce them to drink alcohol.

The rats, however, clearly preferred plain water except on weekends when they would go on real alcoholic binges. This was perplexing at first but it was noted that the automatic time switch on the lights was out of order and the rats were being left in continuous darkness over weekends. Another group of laboratory rats was kept in total darkness without subjecting them to any anxiety stress and their preference also switched from plain water to water with alcohol added. In his article, Dr. Geller refers to this "darkness-induced drinking phenomenon" and relates it to the work reported in 1963 by Nobel Prize winner Dr. Julius Axelrod, who found that the rat pineal gland produced more of the enzyme *melatonin* during the dark nighttime period than when it was light.

Dr. Geller then gave injections of pineal melatonin to rats kept on a regular light-dark cycle and without being subjected to any anxiety. The injections alone turned these rats into alcoholics and Dr. Geller further stated that "it is only through such animal studies that one can hope to attain a clearer understanding and perhaps an ultimate treatment or cure, or both, for alcoholism in humans." If Dr. Geller's findings on "the darkness-induced drinking phenomenon" in rats do ultimately apply to humans, perhaps they might go even further than alcoholism and include other addictions such as drugs. The present findings certainly suggest the need for further research in this direction. To me they point up the increasing recognition of the importance of light in our environment and the role it plays in controlling endocrine functions. Pieces of the mysterious puzzle of the biological influences of light seem to be falling into place, not only in connection with the physiological responses to light, but also touching on psychological ones such as mental attitudes, learning abilities or disabilities and behavioral problems.

With all the accumulating scientific evidence of the biological effects of light, both on the skin and through the eyes, I feel there is justification for further comment on a number of simple observations which I believe indicate the need for more controlled studies. It is my hope that these additional observations may suggest some basis for selection of the parameters to be included in future light studies, and not taken as an attempt to suggest any instant cure for cancer or other human pathologies:

(1) Shortly after the experiment with human cancer patients at Bellevue Medical Center was terminated, a personal friend of mine told me of an acquaintance of his, a man in his early seventies who had just been diagnosed as having terminal lung cancer. This man lived in the southwest and wore sunglasses most of the

time. My friend sent him a copy of *My Ivory Cellar* and a set of the instruction sheets that had been given to the 15 patients in New York City for living outdoors as much as possible and avoiding artificial light sources. The elderly lung cancer patient agreed to follow the instructions. The tumor completely disappeared and he lived for approximately eight years before dying of a heart complication. The diagnosis of lung cancer had been made at a large veterans' hospital, but unfortunately, we were unable to obtain any further medical details regarding the diagnosis.

(2) In 1961, the Communicable Disease Center of the U. S. Public Health Service in Atlanta reported that a school in Niles, Illinois, had the highest rate of leukemia of any school in the country. In fact, it was five times the national average. I made a point of visiting the school and talking with the superintendent, the head maintenance man and also some of the teachers who had been at the school since it was built. I learned that all of the children who developed leukemia had been located in two classrooms, and that the teachers in these particular classrooms customarily kept the large curtains drawn at all times across the windows, because of the intense glare from the extensive use of glass in the construction of the new, modern building. On examining these curtains I found that they were not completely opaque, but more of a translucent type of material that allowed some of the outdoor light to penetrate and which gave them a greenish appearance. With the curtains constantly closed, it was necessary to keep the artificial lights on in these two classrooms, and I learned from the head maintenance man that the original tubes installed were "warm white" fluorescent, which are very strong in the orange-pink part of the spectrum.

After several years of this regular procedure of keeping the curtains closed and the lights on, the class-room teachers of these two particular rooms left the school and their replacements preferred to leave the curtains open and the lights off unless needed. I also learned that at about this time there was a general replacement of the warm white fluorescent tubes and that the new tubes were cool white, which are not as strong in the orange-pink part of the spectrum, and which were not lit continuously, but only as needed. As of the time of my visit in 1964, there had been no further leukemia cases reported for several years. No explanation for the previous unusually high rate of leukemia at this school had ever been found, but the problem no longer existed and the situation had returned to normal.

(3) At a dinner given prior to one of my lectures, I sat next to the daughter of the late Dr. Albert Schweitzer. Our conversation dwelt mostly on her experiences as assistant to her father at Lambarene, on the west coast of Africa. I asked her about the rate of cancer of the people in that area, and she replied that when her father had first started the hospital they found no cancer at all, but that now it was a problem. I asked if the people living there had started installing glass windows and electric lights in their otherwise simple surroundings and she said they had not.

Then I half jokingly asked her if any of the natives wore sunglasses. She looked startled and then told me that the natives paddling their dugout canoes up and down the river in front of the hospital often wore no more than a loin cloth and sunglasses, and indeed, some wore only the sunglasses. She further explained that sunglasses represented a status symbol of civilization and education and had a higher bartering value than beads and other such trinkets. There is, of course, no scientific proof of a correlation between the wearing of sunglasses and cancer, but it does raise an interesting question.

(4) I learned from another elderly acquaintance that he had been diagnosed as having cancer of the prostate, and surgery had been recommended. I found that he had for many years been wearing eyeglasses with a light pink tint, and was able to persuade him to stop wearing these and get some new full-spectrum clear ultraviolet-transmitting spectacles. I also advised him to cut down as much as possible on watching television and spend more time outdoors, or at least on an open screened porch. He has now gone for three years without surgery and the problem has apparently disappeared.

(5) A physician interested in our light research studies advised me that a close friend of his had been diagnosed as having a terminal cancer of a fast-spreading type. This physician advised that under such circumstances, life expectancy of approximately four months was the most that could be hoped for. He told me that while there was no evidence that installing the new fluorescent tubes with the added ultraviolet in his hospital room could do any good, he could not see that any harm could be done. Accordingly, I personally helped to install the fluorescent tubes in the patient's room and later provided some for his room at home when he was able to leave the hospital following surgery. The physician also approved of some additional ultraviolet ocular therapy, with very short exposures to low intensities of short wave ultraviolet produced by a germicidal type lamp, to try to make up the trace amounts of this part of the UV spectrum that are now recognized as penetrating through the atmosphere. Because of the known harmful effects from too much of this short wave ultraviolet radiation, the exposure prescribed was minimal. This patient also continued to

receive chemotherapy. He lived for another ten months and was remarkably active and free of pain during this time.

(6) Another case was that of a friend, a middle-aged woman suffering from a severe case of exophthalmic goiter, or Grave's disease, with severe swelling of the eyes. She had had the maximum of radioactive treatment without responding and had been told by her physician that the chances were that she might lose her sight completely in the near future. She was willing to try anything, even if there was no proof that it might help her, as long as her physician agreed that it could not make the situation worse.

She installed the new full-spectrum fluorescent tubes, with the long wave black light ultraviolet added, in her kitchen and other rooms where she spent most of her time. During the summer she and her husband went to their cottage in the north woods and took the fluorescent tubes with them. That particular summer the weather was cold and rainy most of the time so that she was not often able to get outdoors into the sunlight, which, she said, helped her more than the fluorescent lights did indoors. On her return to Florida in the fall, her physician authorized the added ultraviolet therapy to supplement the full-spectrum fluorescent tubes, with short exposures to low intensities of short wave ultraviolet, again to try to duplicate the trace amounts of these wavelengths in natural outdoor sunlight.

Now, two years later, this woman's condition is greatly improved. The redness, soreness and constant watering of her eyes has stopped. The diplopia, or double vision, has completely disappeared and she can see better and feels better than before the light treatment was begun.

7) The husband of the woman mentioned above had been troubled with skin cancer and on several occa-

sions had undergone minor surgery. He was again having considerable difficulty and his physician had recommended further surgery. However, on his own initiative, he decided to try the same ultraviolet ocular therapy that he was giving his wife for her eye condition, to avoid watching television and to follow the other general instructions. Immediately, his skin cancers began to disappear, and within a matter of four or five months his skin appeared perfectly normal without surgery or other treatment.

(8) Because of the increasing number of reports of vandalism in Sarasota we began to leave our kitchen light on all night as a security measure. The kitchen light happens to be one of the new full-spectrum fluorescent tubes with the added long wave ultraviolet, and the glass in the kitchen window had been replaced with ultraviolet transmitting plastic, as had all the windows and sliding doors in the rooms where we spend most of our time indoors.

Suddenly, I noticed that both an orange tree and a grapefruit tree just outside of the kitchen window were bursting into bloom completely out of season. The blossoms of the orange tree nearest the kitchen window were exceptionally large—in fact, a full two inches across. This was doubly unusual because the lights were on only during the nighttime, and the intensity that reached the trees was extremely low.

A report about this phenomenon in our local newspaper brought a phone call from a registered nurse whose mother had been operated on for a cancer that was considered terminal at the time. After the operation, she brought her mother home. She told me she had placed a small ozone lamp in her mother's room to keep down odors in the sickroom. She said, however, that the small ozone lamp was defective; the light shield was missing, so that the short wave ultraviolet light was shining out into the room.

The location of the light was about fourteen feet from the bed, but as there was a large overstuffed chair in between, it was not possible to determine how much of the time the cancer patient may have been able to see the light directly, and how much of the time she might have received only indirect light reflecting from the walls and ceiling. The light was on continuously during the day and was also used as a night light. She had intended to buy a new shield to replace the missing one, but due to all the confusion at the time she never did. This had happened more than two years ago, and now, her mother, who is in her eighties, is apparently in excellent health and very active for one of her age. A check on this elderly patient's medical record indicated that the type of cancer she had was not of the fast-spreading type, but the circumstances of her recovery still seem to merit investigation.

Observations of this kind must be viewed with extreme caution. On the other hand, I feel strongly that they should not be ignored, especially when so many such coincidences seem to fit into an overall pattern. Further research on the biological implications of light is greatly needed, and I must note, with great satisfaction, that experiments with both laboratory animals and with a limited number of patients under medical supervision are being carried on within the limits of a restricted budget. The results to date are continuing to show clearly an indicated relationship between light environment and tumor development.

Fifteen

ROUTINE OPPOSITION TO NEW IDEAS
AS STANDARD PROCEDURE

No matter how hard one tries to soft-pedal such stories as our turndown by the American Cancer Society, the word spreads, and the fewer actual facts that are released, the greater seem to be the rumors and exaggerations. Several doctors made a point of contacting me to find out exactly what had happened. All seemed amazed and almost indignant that the recommendations of such a distinguished doctor as the Chairman of the Tumor Research Committee of a leading hospital should be so completely ignored without his being asked a single question—especially when his recommendations were supported unanimously by the hospital research committee of other doctors all prominent in their individual disciplines. Some recalled that scientists like Pasteur, Fleming, and Goddard would never have qualified for research funds during the early development of their great discoveries; something must be wrong with our present scientific approach to anything that is new or different from existing conjectures dealing with unsolved problems—problems like cancer. A reexamination of present policies might be in order, in connection with the much publicized appeals for vast sums of money for cancer research in order that "not a single possibility be neglected."

One of the physicians particularly distressed by the American Cancer Society turndown was Charles Galloway, M.D., a staff member at the Evanston Hospital, and an assistant professor of gynecology at Northwestern University Medical School, with which the hospital was affiliated. He telephoned and asked me to come to his office, explaining that one of his patients had a condition of leukoplakia of the cervix, which is generally considered to be an indication of a pre-cancerous condition. He further explained that he had attended one of my lectures, and had read *My Ivory Cellar*. He had told his patient of the experiment with the fifteen patients conducted by the head of cancer research at the hospital in New York City some years previously. He added that his patient wanted to see if getting out into the sunlight more might help her situation.

Dr. Galloway explained that as a result of a routine medical examination back in 1957, he had found that this patient had a definite area of leukoplakia of the cervix and that he had removed it at that time with minor surgery. A new area of leukoplakia very quickly appeared immediately adjacent to the scar tissue. This was cauterized once and removed three additional times, up to 1960, but each time it quickly reappeared. In 1960, he suggested that a complete hysterectomy would not only be advisable but necessary. It was at this point that the patient indicated that she wanted to try getting out into the sunlight more, and Dr. Galloway asked me into his office. That was in February, 1960.

I gave him a set of the same instructions that had been given to the fifteen patients in New York. However, now, as the result of lectures that I had given to the research people of several of the leading plastic companies, new ultraviolet-transmitting lenses were available for eyeglasses, and also ultraviolet-transmitting material was available for use in windows and sliding doors. This patient immediately obtained a pair of the new ultraviolet-transmitting eyeglasses and also had the UVT plastic put in the windows and sliding doors of rooms in her home where she spent most of her time. By May of that year, Dr. Galloway advised me that the area of leukoplakia was definitely smaller, and that he would temporarily, at least, postpone performing the hysterectomy. By October, 1960, the area was still smaller, and at that time he decided to remove it again with minor surgery.

Since then (as of this writing) there has been no reappearance of the condition. Dr. Galloway decided major surgery was no longer indicated and that everything seemed to be perfectly normal. He has given his patient a clean bill of health. Today, twelve years later, this same patient has had no recurrence of the condition, and no further surgery, major or minor, has been necessary. She has also stopped wearing sunglasses.

After the first year had passed with no reappearance of the leukoplakia, Dr. Galloway was sufficiently impressed with the results that he arranged for me to meet Charles Huggins, M.D., head of The Ben May Laboratory for Cancer Research, associated with the University of Chicago. Dr. Huggins, well known for his work in relating tumor development to the malfunction of the pituitary gland, was a recipient of the Nobel Prize several years later. Studies had been underway for some time by Huggins and his staff to determine the effect on cancer in humans through treating the pituitary with drugs, implantations of radioactive beads and by the removal of the pituitary. Removing the pituitary would mean that the patient would have to be given daily dosages of the hormones known to be produced by that gland in the human body.

I attempted to discuss with Dr. Huggins the work of Rowan, Benoit, Assemnacher and others, and the indicated influence of light received through the eyes on the pituitary gland, but was stopped short by his saying that he did not think it possible that light entering the eyes could influence the pituitary. However, as a courtesy to our mutual friend, he took me out to their animal laboratory room and introduced me to Katherine L. Sydnor, M.D., who had been conducting a research project with rats for the past two years. Dr. Sydnor had been experimenting with administering carcinogenic chemicals to the rats and keeping them under different light conditions. Some were also kept in total darkness, and others subjected to various periods of artificial light of different types of both incandescent and fluorescent. Some of the rats had been blinded and the eyes actually removed and the eyelids sewn together. Dr. Sydnor was just completing this project and planning to leave the university within a matter of a few days for a year's study at a cancer research institute in London, England.

My visit with her was brief, but she showed me the rats and remarked in particular that the fur of those kept under total darkness was soft and smooth in texture, and also quite thick and fully developed. The rats exposed to the artificial lights were of the same breed but looked entirely different; the fur was coarse and extremely bristly. A good many of the animals under the daylight white fluorescent light were completely bald on the tops of their heads and, in many instances, this baldness continued down the ridge of the back. Dr. Sydnor advised that the rats would be sacrificed within the next few days and promised to call me before leaving for London if any significant results were observed in the rats at autopsy. I received a phone call and she told me that the tumor development in the rats kept under the lights was significantly greater than in those kept in total darkness, and that those rats kept under the regular incandescent light showed considerably more and larger tumor development than the rats under daylight white fluorescent.

I asked if this information would be published. The answer was that the results came as such a complete surprise that the experiment would have to be repeated several times before publication of anything so startling could take place. Since she was leaving for England, it was uncertain when or if the experiment could be repeated.

This additional bit of evidence that light influences tumor growth certainly added support to the overall picture. Each of the several individual experiments by themselves had not aroused much general interest, but if all the results could be put together in one combined report and submitted to the various cancer societies and research centers, some recognition or further progress might be made.

Some years prior I had been invited to show the cancer cell pictures made for Northwestern University Medical School and the pollen germination sequences, along with others of the earlier time-lapse films, to a group of doctors on the regular staff at the American Medical Association headquarters. It was my understanding at the time that they were from different departments within the organization and together formed a research reviewing committee. Shortly after my meeting with them, a six-page article appeared in the November, 1959, issue of *Today's Health*. It was called *Meet John Ott*. It reported nearly everything I had said at the meeting and included illustrations of the cancer cells dividing, pollen grain activity, and even the pumpkin sequence from *My Ivory Cellar*.

After some years, I was invited again to show my films and this time in the Board of Directors' room at A.M.A. headquarters. During this time I had accumulated quite a file of correspondence and had made a number of personal visits for the purpose of discussing various matters regarding our light research studies. It therefore seemed quite logical to suggest presenting this latest combined information of the results of all the experiments that had been carried on at the different hospitals and medical centers to the American Medical Association. I did, and very quickly received a reply from the Director of the Division of Scientific Activities, as follows:

March 2, 1964

Dear Dr. Ott:

Dr. Blasingame has requested that I reply to your letters of February 13 and February 19, 1964. We appreciate that you have supplied us with the information that accompanied your letters.

You suggested that you might like to "report the latest information on (your) current mouse experiments to (our) research committee." The American Medical Association does not have a "research committee," and I would suggest that you continue to divulge the results of your experiments through normal channels of scientific communication.

Sincerely,
(signed) Hugh H. Hussey, M.D.,
Director
Division of Scientific Activities

Not long after the receipt of the above letter, I was surprised to be invited by both of the co-chairmen of the A.I.A. (American Institute of Architects)-A.M.A. Joint Committee on Environmental Health to show the time-lapse films and present our whole story again in the Board of Directors' room at A.M.A. headquarters and again I quote from another, more encouraging letter as follows:

December 27, 1965

Dear Dr. Ott:

On behalf of the Joint AIA-AMA Committee on Environmental Health, I want to thank you for the most stimulating and highly informative presentation you gave on December 3 on the effects of artificial lighting on plants and animals and the implications these may

179

have for environmental health....Your imaginative findings suggest that the effects of illumination on human health and well-being should be further explored. This is certainly an area of direct interest to the Joint AIA-AMA Committee on Environmental Health....

> Sincerely,
> James H. Sterner, M.D.
> Co-Chairman
> Joint AIA-AMA Committee on
> Environmental Health

I received several other rather complimentary letters regarding our light studies from various members of the same Committee and especially from Edward Matthei, the other Co-Chairman representing the A.I.A. Mr. Matthei was a partner in the well-known architectural firm of Perkins and Will. I learned that the A.I.A./A.M.A. Committee had developed plans for a World Light Symposium, and I was asked to prepare and present a paper on the results of all of our light studies. This seemed to be the perfect opportunity to write a report combining the results gathered from all the doctors we had worked with. I submitted such a report to Drs. Wright, Gabby, Scanlon, Galloway, and Sydnor, the principal investigators of each of the five major experiments.

What was most gratifying to me was the willingness of each of them to co-sign the overall report personally. I thought any report bearing the signatures of five such prominent scientists from recognized medical institutions would certainly draw the attention of others to the need for further studies of the biological effects of light on animals, and more specifically to the possible relationship between light and tumor development. In submitting this report to the A.I.A./A.M.A. Joint Committee, I also suggested that I might include it in a

180

paper I was scheduled to present to the annual meeting of the American Association for Laboratory Animal Science to be held in Las Vegas, Nevada, in October 1968. By so doing, the paper might be published, and therefore be more available as a reference. This plan seemed a good avenue of approach, until the following letter was received:

October 18, 1968

Dear John:

Pursuant to your letter of October 10, I have read the paper which you will present at the forthcoming AALAS meeting in Las Vegas. I am afraid that I cannot comment on your suggestions that light directly affects the gonads and influences the retinal-hypothalamic-endocrine system as such matters are beyond the scope of my knowledge.

I recently came across two items which may be of interest to you. Some work by Ludvigh and McCarthy (*"Absorption of Visible Light by Refractive Media of the Human Eye," Arch. Opth.* 20:37, 1938) and Kinsey (*"Spectral Transmission of the Eye to Ultraviolet Radiation," Arch. Opth. 39*: 508, 1948) suggest that ultraviolet radiation does not penetrate the eye, which is in direct contradiction to your thesis. Also a recent paper by Daniels, *et al. (Scientific American* 219, No. 1, July, 1968) reviews some of the harmful effects of u.v. radiation to the skin.

The AIA/AMA Joint Committee has decided to hold the matter of a light symposium in abeyance until more scientifically-oriented material is available.

I trust this information will be helpful.

Sincerely,
Gordon R. Engelbretson, Ph.D.
Assistant Director
(Department of Environmental Health,
American Medical Association)

Again, what seemed like such a good move toward a world light symposium turned out to be another blind alley, and I was further disappointed to learn shortly thereafter that the Joint Committee on Environmental Health of the A.I.A./A.M.A. had been terminated.

Fortunately, there are ways out of blind alleys, and this was not the first time I had had to back up and try a different route. This time it seemed advisable to back up quite a distance, to stay on the more heavily traveled highway and not venture so far from my home base of time-lapse photography.

While doing this I had plenty of time to obtain copies of the articles mentioned in Dr. Engelbretson's letter. After reading them all carefully, it did not seem that what they said was at all in direct contradiction to my own thesis.

The article by Ludvigh and McCarthy, *"Absorption of Visible Light by the Refractive Media of the Human Eye,"* deals only with the absorption of visible light by the refractive media of the human eye and, as more specifically stated in the article, covers only the range of wavelengths from 400 to 820 millimicrons. The paper further states that many investigations have been made of the absorption of infrared and ultraviolet rays by others, but since the literature indicated that there were no adequate data available on the absorption spectrum of the ocular media for light of visible wavelengths, it was decided that these data should be obtained. I cannot see that this paper in any way contradicts my thesis regarding ultraviolet.

On studying the second paper entitled *"Spectral Transmission of the Eye to Ultraviolet Radiations,"* by V. Everett Kinsey, Ph.D., I quote as follows: "the cornea and aqueous and vitreous humors absorb most of the radiations of wavelengths shorter than 300 millimicrons, (3,000 A.) i.e., the abiotic portion of the ultraviolet spectrum. The corneal epithelium absorbs chiefly radiations of wavelengths shorter than 290 millimicrons (2,900 A.)," which is exactly where the atmosphere cuts out the shorter ultraviolet wavelengths of sunlight. The charts included in this article show that much of the ultraviolet down to these specific wavelengths is transmitted by the various parts of the eye. In other words, the longer wavelengths that penetrate the atmosphere also penetrate the eye.

After studying this article, I find no contradiction between what Dr. Kinsey states and my own thesis on ultraviolet. I do think there is a need for a better general understanding of the different intensities of near and far ultraviolet in natural sunlight as compared to the equivalent wavelength intensities produced by many artificial ultraviolet sources.

Ludvigh and McCarthy state in their paper, "the absorption characteristics of the human crystalline lens, for light of visible wavelengths, change in a fairly regular fashion with age," and that "changes in the infrared and ultraviolet absorption spectrums with age have been noted." Kinsey stresses in his paper that all measurements of ultraviolet absorption were made using eyes obtained from freshly killed young rabbits, which I would infer as indicating the ultraviolet absorption spectrum might be different in the eyes of older rabbits. The ultraviolet absorption figures made by both Ludvigh and McCarthy and also by Kinsey were made using "dead" eyes. If the age of an animal can make a difference in the transmission characteristics of the various tissues of the eye, then it seems to me there might be even a greater difference in the ultraviolet transmission characteristics of the various tissues of the eye depending on whether they were dead tissues or were part of a whole living eye, intact and receiving its normal *supply* of blood and nervous energy through the many intricate connections with the central nervous system, as well as the many nerve fibers connected to the accessory or autonomic nervous system.

An article in the July 1968 issue of *Scientific American* was entitled *"Sunburn"* and written by Farrington Daniels, Jr., Jan C. van der Leun and Brian E. Johnson. This excellent article discusses what happens to the skin from excessive or over exposure to sunlight and dwells especially on the sensitivity of people (who have been working indoors for long periods of time) when exposed directly to the sun.

The authors of the article then state that they have deliberately left a discussion of the beneficial effects of sunshine for the conclusion of the article because the constructive aspects of this radiation surely outweigh the destructive ones for both animal and plant life. The article states that the best known specific benefit of exposure of the skin to the sun is the production of vitamin D. This vitamin is essential for the absorption and metabolism of calcium, and a deficiency of the vitamin in growing children results in the bone disease called Rickets. The authors further state that basking in the sun confers many benefits besides the synthesis of vitamin D—some that are known and undoubtedly many others that have not yet been explored. The article also points out the effect of sunlight on the skin surface—that it prevents bacterial and fungal infections. It mentions that studies in Northern Europe and the U.S.S.R. have shown that indoor workers and those in lands weak in sunshine improve in physical fitness when given moderate supplementary doses of ultraviolet radiation.

So much for the items in the Engelbretson letter.

Sixteen

SIGNS OF ENCOURAGEMENT

Further research is certainly needed, but during 1968, the Federal Government began to cut back on funds for research. The space program was ultimately curtailed drastically. Thousands of scientists became unemployed. This started a chain reaction as competition increased for what funds remained. During 1969 and 1970, support for new projects was increasingly difficult to obtain when those already existing were being so severely slashed.

Corporate profits were down due to the business recession. The stock market was in a major slump. Interest rates rose sharply. Money was tight, and trying to raise funds for a study involving the effects of pink light on the development of cancer was next to impossible.

Some of the reasons given for turning down our various requests for grants were interesting. One foundation offered funds for assisting young scientists under forty years of age. Tony, my assistant, was forty-one at the time. Other foundations, particularly interested in cancer research, found that while our proposal was of an intriguing and potentially significant nature, it did not come within the scope of their present programs. Several of the largest foundations quite frankly stated that the constraints on their budgets were so great that it was impossible for them to step outside their existing program priorities to make even a small grant.

Many others said a polite "no," along with extending best wishes that we might be able to obtain funds elsewhere. Hopes for expanding or even continuing our limited program were more in doubt with each rejection. Attempts to persuade those in charge of existing research studies to change some of their light bulbs and see if they could then observe any difference in their research findings were to no avail, even when we offered to furnish the light bulbs.

Still, more and more requests for the time-lapse films kept coming to me from medical and scientific groups all over the world. One lecture-film trip during September, 1971, was of particular interest. The first stop was at the Western Regional Research Center in Berkeley, California, to give a seminar and show the time-lapse films to a joint meeting of U.S.D.A. and N.A.S.A. scientists working on specifications for the proposed first space station. Their particular concern at the time was in designing a lighting system for growing vegetables for the astronauts in outer space. The scientist in charge of this phase of the program told me he remembered visiting my greenhouse in Lake Bluff, Illinois, many years before, and had decided to ask me to come out there after seeing the time-lapse pictures of roses, geraniums and other flowers in the film, *On A Clear Day You Can See Forever.*

My next lecture was to a group of researchers at the National Marine Fisheries Service Research Center at Auke Bay, Alaska. Then on to Decatur, Illinois, to discuss a problem concerning one of the most valuable pure bred Arabian stallions in the world. This stallion had for some time been siring approximately 90 per cent male colts. The problem resembled that of the chinchilla breeder in New Jersey, where changing the lights in the chinchilla breeding rooms had solved the difficulty. But would light then do the same for horses? It seemed worth trying to find out whether changing the lights in the stable might result in breeding more fillies, which were so badly needed. Accordingly, another experiment was started to test what effect changing the lighting environment might have on the sex of the progeny of this valuable Arabian stallion. It will take at least a year to determine the results.

The next lecture and film showing was in Coldwater, Michigan, at the annual convention of the Federal Organic Clubs of Michigan. Then two more lectures, including the time-lapse films, at the University of Vermont in Burlington. One was for the Department of Community Medicine, where a great deal of emphasis is now being placed on studies of environmental problems under the guidance of Dr. Charles Houston. The second was at the affiliated Mary Fletcher Hospital, where Dr. Jerold Lucey and others are doing outstanding work treating jaundice in premature babies by light therapy.

On the way back to Sarasota I stopped briefly in New York City and then at the Wills Eye Hospital in Philadelphia, where another study of the effect of light on mice was being started; again supported by the Wood Foundation. After three days at home in Sarasota it was time to start out again on the next lecture trip, back to Vermont, where I showed the time-lapse films to the annual convention of the New England States Natural Food Associates being held in Manchester. September, 1971, was a busy month.

Many people have gone out of their way to pass along information on the biological effects of light to others engaged in cancer research. As a result, I met Edward C. Delafield, who formerly had lived in New York City and was now a resident of Sarasota. Mr. Delafield had for many years served as Treasurer of the Memorial Hospital and Trustee of the Sloan-Kettering Institute for Cancer Research. Mrs. Delafield and other members of the family had been active in raising funds and otherwise helping in making Sloan-Kettering the outstanding cancer research institute that it is today. I've never known anyone more interested in cancer research and eager to do something about it than Ed Delafield. On January 20, 1971, he wrote a letter to the director of research at Sloan-Kettering, Dr. Frank Horsfal, from which I quote in part as follows:

> I have a friend down here named John Ott (Sc.D., Honorary) who has been working with light for many years. I understand that you and Mrs. Ott have known each other for many years and that at an early stage of his experimental work, Mr. Ott sent you a paper on it. Since then, he has carried it much farther and I feel that his findings merit your attention, as he has come across some wonderful things in connection with the possibility of curing cancer.
>
> I am enclosing a summary of his work along this line only, and do not include the rest of his experiments with light and radiation. I think you will be interested in this memorandum and I am sure Mr. Ott will be glad to come to see you if you think it would be personally helpful. I am also enclosing a copy of his own writing paper, on which you will note a list of the doctors down here who are now acting as Medical and Scientific Advisors to the Light Institute directed by Mr. Ott. You will also note the name of Dr. Phyllis Stephenson, who was with you a few years ago, and who is now practicing medical oncology here in Sarasota. She is also interested in how much the ultraviolet rays can help over a long period of time.
>
> After you have read this memorandum let me know your thoughts . . .
>
> Sincerely,
(signed)
Edward C. Delafield

The following reply arrived shortly:

Dear Mr. Delafield:

We had hoped Frank would be able to respond to your letters of January 20 and February 5, but this has not been possible. Therefore, with the aid of one of our biophysicists, I am responding to your query about light and cancer.

We both enjoyed reading the brochure and can only comment that the observations have been recorded with great care. The author, however, has given no evidence of controlled experimentation which is required to establish the validity of the claims made. We are unaware of any publications by others attempting to evaluate the influence of light on cancer nor do we know of others who have attempted such studies.

Two incidental bits of information may be of interest to you. One, increasing interest has been manifest regarding the pineal gland. Although this is solidly encased in the skull, it responds in some way to light. The gland seems to have an as yet undefined role in the control of the endocrine glands, some of which are known to affect the progress of neo-plastic* disease. More questions are posed than answers provided by our present knowledge of this gland.

Another interesting observation is that almost all therapeutic agents for cancer (radiation, chemical substances, etc.) are cancer-producing under some circumstances. Since the association of sunlight and the appearance of skin cancer is well established, it would not be surprising if light proved to have some therapeutic effect. This may be very difficult to establish.

Sincerely,
Leo Wade, M.D.,
Vice-President and Deputy Director

The reason Dr. Horsfal was unable to reply in person was that he had just been diagnosed as having terminal cancer of a rapidly spreading type. Dr. Horsfal died shortly thereafter, and his death was a great loss, not only to Sloan-Kettering but to the whole world of cancer research. The President of Memorial Hospital might have answered Ed Delafield's letter had he not also been suffering from a terminal cancer. His death was

* Any abnormal growth such as cancer.

a double blow to Sloan-Kettering. That two of the most prominent and respected men in the field of cancer research should both die of cancer was tremendously disheartening. Still, cancer research must go on, in spite of such setbacks, so that more can be learned about what causes cancer and possible ways to cure or prevent it.

To me, the information in the reply to Ed Delafield's letter is tremendously exciting, and further suggests a possible relationship between light and cancer.

In an article entitled *"Fashions in Cancer Research,"* published in *The Year Book of Pathology and Clinical Pathology,* 1959-1960, Dr. Robert Schrek, Assistant Professor of Pathology at Northwestern University School of Medicine, discusses many types of cancer research that have been popular from time to time. With the invention of the microscope, scientists began looking for microbes or germs as a possible cause of cancer. Then, looking for parasites became popular. Various organisms were reported as suspected causes of cancer. Theories have changed from time to time on the question of whether cancer might be contagious or hereditary. Certain types of cancer have been related to smoking and various occupational hazards. Today, the virus theory is very much in fashion. Dr. Schrek further points out that human nature is the same whether in the field of science, medicine, business or art. He states that we have, will have, and should have fashions in cancer research. His article tells how some of the important developments in the early days of cancer research were either ridiculed or completely ignored at the time of their discovery, and he goes on to say that cancer research is wedded to fashion, for better or for worse. "One of these may yield a cure for cancer, but who knows—the final answer may not be within the limits of current fashions in cancer research."

Periods of discouragement are unquestionably a part of all research. During such periods, anything the least bit encouraging is greatly appreciated. In the past, a little encouragement always seemed to come from somewhere, frequently from the most unsuspected places. Such was the case with one of our films, *Gateway to Health,* produced for the International Apple Institute. The story dealt basically with the importance of proper nutrition in maintaining not only good general health but especially healthy teeth. Needless to say, it stressed the need for apples in the diet, along with plenty of other fresh fruits, vegetables, dairy products, meat and fish, etc.

What made the film so convincing was the knowledge and sincerity of Dr. Fred Miller, of Altoona, Pennsylvania, who played the leading role in a story about a remarkable dentist—himself. Dr. Miller was not a professional actor, nor had he had experience as an amateur one, but he told his story about taking care of his patients' teeth in a way that has sold over 1000 copies of the film to schools, civic clubs and organizations interested in the importance of nutrition and how to keep teeth in a strong, healthy condition.

Dr. Miller became interested in the possible biological effects of light on animals; the need for natural sunlight seemed to tie in closely with his ideas about the importance of natural foods that had not lost much of their vitamin and nutritional content through modern methods of processing and preserving. He invited me to attend several dental society meetings, including an early morning breakfast in Chicago on February 19, 1962. It was at this meeting that Dr. Miller introduced me to his friend, Dr. James Winston Benfield, a dentist from New York City. Dr. Benfield was vitally interested in the subject of nutrition, and had also become interested in the study of the biological effects of light on animals, particularly the human animal.

At that time so many new developments were taking place and so much going on that I asked Dr. Benfield to verify the chronological order of our combined efforts to develop the first full-spectrum fluorescent tube and the starting of further important research with this new source of artificial light. From his reply I quote as follows:

October 22, 1971

Dear John:

The copy of my letter to you dated March 25, 1962, that you sent to me recently definitely establishes the date of our first meeting as February 19, 1962. At that time, you had written a paper for the *Illinois State Medical Journal* and Dr. Fred Miller introduced you to the then editor of the *Journal of the American Dental Association* in my presence. We were both guests of Dr. Miller at a breakfast meeting at the Blackstone Hotel in Chicago. The editor of the *ADA Journal* did not seem interested in learning anything about your paper.

However, having overheard the conversation, I asked you for some of the details about your work, and that was my introduction to the biological effects of light. Subsequently, Fred Miller and I came to Philadelphia to hear you speak at the Wills Eye Hospital. In 1965, I was a member of the program committee of the American Academy of Restorative Dentistry. In February of that year I proposed that you be put on our 1966 program. When the suggestion was approved, I came out to your Lake Bluff, Illinois, laboratory to personally invite you to do so.

I was still a member of the program committee in 1966, and, at our meeting following your presentation to our Academy in February of that year, there was great enthusiasm for your paper. The suggestion was made by other members of the committee that the Academy should have a progress report on the subject of light in the 1967 program, and for want of another speaker on the subject, I was asked to give that report. The members of the Academy had sensed the importance of your findings, because they, as dentists, probably spend more time under artificial light sources than those in almost any other profession.

During the summer of 1966, when I was preparing material for the 1967 meeting, I decided to try to interest one of the light manufacturing companies in producing a fluorescent light that would duplicate daylight. At that time you were recommending the use of black light fluorescents as a supplement to standard fluorescents, but this, obviously, was not the most practical means of providing the near ultraviolet portion of the spectrum. You had explained that you had previously tried to get first General Electric and then Sylvania interested in producing a full spectrum light source without success. The idea had been vetoed by the medical department of General Electric. Although I had good contacts with both of these companies through my brother, who was a distributor for both General Electric and Sylvania, I decided that, in view of your experience, there was no point in approaching them. At about that time, I read in *Science News* that the Duro-Test Company, the smallest of the four manufacturers of light sources in the U.S., had brought out the Optima tube for color matching in the textile, dye and printing industries. That gave me the idea of approaching them, since they had gone part way in the right direction. I wrote a letter to the President of Duro-Test and was about to mail it when I happened to discuss the subject of light with Mr. Alexander Imbrici, who has been a patient of mine for many years. He seemed particularly interested in learning about its biological effects, so I went on to tell him about the letter I was about to send requesting an appointment with one of the manufacturers. He asked what company I was writing to, and when I told him it was the Duro-Test Co., he said, "Don't bother to write the letter, just let me handle it for you." I had not known until then that he was a personal friend of the president of the company and did not learn until much later that he is one of its largest stockholders.

In a few days, he had set up an appointment for me with several of the Duro-Test executives at their Light

Bulb Center in New York City. I spent several hours with them and told them many things about the biological effects of light that they had never heard of. They said they had had a few complaints from people working under fluorescent lighting that it caused headaches and even some strange complaints that it causes a change in the menstrual cycles in women. They thought this was all nonsense. However, after the evidence I presented, they began to see that perhaps there was something to the idea that their products could be improved if they were made to reproduce daylight. They invited me to visit their North Bergen, New Jersey, manufacturing plant.

I spent a day there with members of their research department. They took me through the plant and showed me the manufacturing operation. Their first reaction to my recommendation was that it was not possible to fabricate a fluorescent tube that would precisely duplicate daylight, but they agreed to try—and to my recommendation that they retain you as a consultant. Approximately six months later the problem had been solved and plans were being made to market the product.

In my 1967 paper to the Restorative Academy, I stated that Dorland's Medical Dictionary refers to near ultraviolet as the *vital rays*. Up to that time, Duro-Test had called their new light source "Optima FS." A copy of this paper was sent to Duro-Test's public relations man, which contained the quote from Dorland's Dictionary. Shortly thereafter, the name was changed to Vita-Lite. I have always thought that this was probably the basis for the change in the name, although they have never said so.

This is the story of the birth of the first full-spectrum fluorescent light as it actually occurred. Fred Miller was instrumental in introducing you to me, and the Academy was responsible for asking you to speak to them in 1966 and for asking me to give the 1967 progress reports. If it had not been for that chain of events, there probably would not be a full-spectrum light available today.

Sincerely,
James W. Benfield, D.D.S.

194

I served as a consultant to the Duro-Test Corporation for approximately three years, and it was during this time that their new full-spectrum tubes were developed and put on the market.

When Dr. Benfield was elected President of the Samuel J. and Evelyn L. Wood Foundation, Inc., he invited me to show the time-lapse pictures to the Trustees of the Foundation at a dinner in his home. Shortly thereafter, the Environmental Health and Light Research Institute received a grant from the Wood Foundation that made possible the project at the Stritch Medical School of Loyola University in Chicago under the direction of Dr. Alexander Friedman, head of the Department of Pharmacology.

Further encouragement came from Dr. Lewis W. Mayron at the Veterans' Administration's Edward Hines Jr. Hospital, which is adjacent to the Stritch Medical School:

Dear Dr. Ott:

With reference to your articles in *My Ivory Cellar* and the November, 1959, issue of *Today's Health*, concerning ragweed pollen and the effect of nasal secretions from an allergic individual, I should like to cite some preliminary experiments that I have performed in my laboratory.

The mixing of ragweed pollen and the saliva of "nonallergic" individuals differs quantitatively from when "allergic" saliva is used, in producing droplet formation and, in some cases, stalk formation, on the pollen grains. Results with saline are similar to those obtained from nonallergic saliva.

I wish to use this reaction as an assay for studying the chemical and biochemical reactions involved. However, I do not have time-lapse photographic equipment. Would you be interested in collaborating on such a project?

I have not yet made application for funding for this work because I want to be sure I could repeat your initial observation before so doing. However, I am now ready to apply for funds, and will start writing a proposal. In view of the increasing difficulty in obtaining federal funds, can you suggest any funding sources?

I eagerly await your reply.

Sincerely,

Lewis W. Mayron, Ph.D.,

Supervisory Research Chemist,

Dental Research

The fact that Dr. Mayron's study was undertaken independently of mine and showed similar results seemed to be of considerable significance.

I contacted some of the larger pharmaceutical companies again, but unfortunately financial assistance was not possible, due to cut-backs in each company's research budget. The project might have died on the vine had it not again been for Dr. Benfield's intervention on our behalf. He discussed the possible significance of Dr. Mayron's findings with the other trustees of the Wood Foundation, and the result was another grant to the Environmental Health and Light Research Institute to support further studies of the reaction of ragweed pollen. However, shortly after the studies were underway, Dr. Mayron was advised that general budget cut-backs at the Hines Hospital would necessitate the closing of his laboratory. Our joint ragweed study was thus interrupted until other suitable laboratory facilities can be made available. Delays are always disappointing, but the significance of Dr. Mayron's findings remain undiminished in their implications.

Although the ragweed pollen project would have to wait for the time being, interest in our TV radiation studies began to expand. Dr. Dickinson contacted the superintendent of the Sarasota County school system, and several meetings were held with its various representatives. The time-lapse pictures always proved to be of special interest. Methods were discussed for studying the possible effects of light and radiation on both the learning abilities and disabilities and the various behavioral problems of school children.

The Sarasota County school system had recently set up a special facility at the Gulf Gate school where children with such problems were being sent for special care. It was known as the Adjustive Educational Center. Mrs. Arnold C. Tackett, the principal, asked me to repeat the showing of the time-lapse pictures at one of the regular Parent-Teacher meetings. She was especially anxious for me to include the pictures showing the effects on the bean plants and white rats placed close in front of a TV set. This meeting was held on the evening of April 27, 1971.

All the teachers and parents were genuinely concerned about the effects possible radiation from TV sets might be having on their children. A plan was worked out for testing the TV sets in each of the homes where the children spent many hours watching their favorite programs. This TV study was also made possible by the continued generous support of the Wood Foundation, and also by the public through a membership drive. Many individuals interested in this radiation problem made contributions by becoming members of the Environmental Health and Light Research Institute. In addition to supporting further research, each membership entitles the subscriber to receive the EHLRI News Letter for one year. This has been initiated to keep all members abreast of what is going on of interest at the Institute.

The results of this first pilot study were no less than electrifying. Measurable amounts of X-radiation were found in all sets that had not been recently repaired, and even some that had been fixed were not perfect. I was invited to speak and show the films at one of the regular dinner meetings of the Manatee-Sarasota Radio

& TV Dealers Association. I thought of how Daniel must have felt in the lion's den, but was soon overwhelmed by their genuine interest and concern in the problem of X-radiation.

All the defective sets that the children at the Adjustive Education Center were watching at home were either repaired or discarded. The location of sets was rearranged, so that none would back up against a wall where anyone might be working or sleeping in the next room. Parents cooperated in making their children sit back as far as possible and by restricting the number of hours the children could watch TV. During the summer vacation, a greater effort was made to interest the children in more outdoor activity.

On November 12, 1971, approximately two months after school had resumed after summer vacation, Mrs. Tackett advised me that an improvement had been noted at the school in the behavioral problems of the group of children in whose homes we had found TV sets giving off excessive amounts of X-radiation. She noted in particular that the two most hyperactive children had been transferred back to their regular school and were acting normally and getting along fine in their classes. One of these was a little girl who had been sleeping on the other side of the wall from a TV set which we found had been giving off radiation from its back. The amount of radiation that we measured on the other side of the wall was .3 mrh. This is within the "safety" standard of .5 mrh as set up by the 1968 Radiation Control Act.

Ben Funk's Associated Press story of April 24, 1970, which I quoted earlier, says:

> The standards, to be fully applied June 1, 1971, require that no TV set may spill out more than 0.5 Milliroentgen of radiation per hour—a level considered safe

at the time the standards were drafted.

However, recent findings by scientists in the Department of Health, Education and Welfare (HEW) indicate that X-ray emissions below the 0.5 level and on down to zero penetrate body tissues with subtle but harmful effect.

"The only answer to the problem," says Dr. Arthur Lazell, assistant director of HEW's Bureau of Radiological Health (BRH) is to "eliminate radiation entirely from the receivers."

The 1968 Radiation Control Act was a big step in the right direction in attempting to establish a safety factor for X-radiation exposure. However, it must be remembered that the 1968 standards represented the ninth time that these standards have been lowered on what seems to have been not much more than a "by guess and by golly" basis each time.

During the mid-1960s, the theme in a great deal of TV advertising was which company had the brightest picture tube. It was common practice for the TV repair men to turn up the high voltage regulator to make the TV pictures even brighter. During this period our country experienced riots in some of our largest cities. Planned deliberate confrontations with police and other law enforcement agencies reached a peak. The disorder during the 1968 Democratic National Convention in Chicago, the fire bombings of department stores along downtown State Street, and the destruction of property along the near northside residential area will not soon be forgotten.

Some of these violent actions have tapered off along with the educational programs set up by the TV industry to warn its personnel in sales agencies and repair shops of the radiation hazard problem, the need for careful checking and the avoidance of stepping up the high voltage beyond recommended specifications. This

is another forward step, but nevertheless, the basic crime rate in the country, according to the FBI annual reports, has continued to show a steady increase. The practice of administering behavioral modification drugs to school children has also been increasing at what to me seems to be an alarming rate.

In 1950, the United States had the fifth lowest infant mortality rate in the world, but eighteen years later we had dropped to thirteenth place. Dr. Jean Mayer, nutrition advisor to President Nixon, has pointed out that the U.S. ranks thirty-seventh among nations as to the life expectancy of twenty-year-old men, and twenty-second for women of the same age. Something is obviously causing an alarming deterioration of the national health record of this country, and this may lead to disaster if the trend is not soon reversed.

As this book goes to press, it is gratifying that more and more medical and scientific research studies are dealing with the biological effects of light. It is my sincere hope that some of the simple observations made possible through time-lapse photography may have been helpful in stimulating further interest in this important subject.

For example, the following letter just received from the Wills Eye Hospital is very encouraging:

Dear John:

Our preliminary experiments dealing with the effects of light on animal tumors are still in progress but we do have some encouraging results. It is still too early to express any formal conclusions because we have had to devise a suitable model for testing your hypothesis.

We have been working with the Harding-Passey malignant melanoma in BALB-C mice and our latest results indicate that mice kept under simulated daylight as compared to cool white fluorescent light develop tumors at a slower and diminished rate. We feel that the work deserves repetition and additional variations in experimental design. We plan to determine if the above-mentioned results are valid and, if so, if the light exerts its effects via the visual tract or via the skin. We are also looking to you for additional experimental light sources as we have discussed and are prepared to test these as soon as they become available.

I will keep you informed with the progress of the work.

With best regards,
Theodore W. Sery, Ph.D.,
Director of Research
Wills Eye Hospital
and Research Institute

These results, though still of a very preliminary nature, do nevertheless seem to tie in with those previously reported by Drs. Wright, Gabby, Scanlon, Galloway and Sydnor mentioned earlier in this book.

Another most encouraging last minute development has been the formation of the American Society for Photobiology and the development of plans for its first scientific meeting to be held June 10-15, 1973, here in Sarasota, Florida. Dr. Kendric C. Smith of the Department of Radiology at Stanford University Medical Center has been elected President and the governing councilors are all scientists highly regarded in their individual disciplines. The formation of the Society is certainly a great step forward and will now furnish the necessary scientific credibility to future studies of the effect of light on man's general health and well-being. The dawn of a new era in research is at hand.

AFTERWORD

In the three years since this book was first published much has happened in the field of research on the effects of light on our health. We have advanced our knowledge on light and cancer and on light and abnormal behavior such as depression and hyperactivity. We have learned more about the effects of combinations of drugs and light on the chemical balance of the body. At this time, a Chicago-based company has begun the manufacture of full-spectrum fluorescent light bulbs for public use.

The following gives some of the highlights of these new developments in the field of health and light.

During the first five months of 1973, a pilot project conducted by the Environmental Health and Light Research Institute in four first-grade classrooms in a windowless school in Sarasota, Florida, showed dramatic reactions in hyperactive children.

In two of the rooms, the standard cool white fluorescent tubes and fixtures with solid plastic diffusers remained unchanged. The plastic diffusers in these fixtures stopped the transmission of any trace of long-wavelength ultraviolet.

In the other two rooms, the cool white tubes were replaced with new, full-spectrum fluorescent tubes that more closely duplicated natural daylight. Lead foil shields stopped any trace amounts of radiation from the cathodes. (This is the same kind of radiation found in TV picture tubes or X-ray machines, but at lower voltages.)

A combination aluminum "egg crate" and wire grid screen, in addition to allowing the full-spectrum light to pass through unfiltered, grounded the radio-frequency energy given off by all fluorescent tubes. This radio-frequency energy is known to cause inaccurate readings from the very sensitive equipment used in the scanning rooms of hospitals and also from some computers. A Russian paper reports that the radio-frequency energy from fluorescent tubes was recorded in EEG readings of human brain waves.

Time-lapse cameras were concealed in specially constructed compartments to take sequences of pictures randomly. The teachers knew of the program, but not when the pictures would be taken. The children were unaware that any pictures were being taken. Time-lapse pictures of the children were made just to see what, if anything, of interest might show up; and something certainly did.

Under the standard cool white fluorescent lighting, some first-graders demonstrated nervous fatigue, irritability, lapses of attention, and hyperactive behavior. Within a week after the new lights were installed, a marked improvement in the children's behavior began to appear.

Without any use of drugs, the first-graders settled down and paid more attention to their teachers. Nervousness diminished, and teachers also reported that overall classroom performance improved. In the rooms with the standard cool white, unshielded lighting still in operation, students could be observed fidgeting to an extreme degree, leaping from their seats, flailing their arms, and paying little attention to their teachers.

In the rooms with the full-spectrum shielded lighting, the same children were filmed once a month for four more months. Behavior was entirely different. Youngsters appeared calmer and far more interested in their work. One little boy who had stood out in the first films because of his constant motion and his inattention to everything had changed to a quieter child, able to sit still and concentrate on classroom routine. According to his teacher, he was capable of doing independent study and had even learned to read during this short period of time.

Similar results were reported in experiments conducted at two schools in California. And an extension of the hyperactivity study by a group of eight dentists, all members of the Sarasota County Dental Society, showed a very significant difference in the number of cavities and in the extent of tooth decay in the new teeth (six-year molars) of the children under the radiation-shielded, full-spectrum fluorescent lights. The improved lighting resulted in one-third the number of cavities, which correlates well with the results of similar experiments performed with laboratory animals (as reported by S. M. Sharon, R. P. Feller and S. W. Burney).

There are undoubtedly many factors contributing to hyperactivity and learning disabilities. However, these observations clearly indicate that light and radiation may be additional stress factors that must be considered.

The fact that no drugs were used is of particular significance, for warnings are now being heard about the widespread use of amphetamines and other psychoactive drugs on children thought to be hyperactive. As child psychiatrist Dr. Mark Stewart of the University of Iowa pointed out (*Time,* February 26, 1973), the danger is that "by the time a child on drugs reaches puberty, he does not know what his undrugged personality is."

Estimates of the number of children in this country now taking drugs range as high as one million. This situation has prompted the Committee on Drugs of the American Academy of Pediatrics to propose regulations to the U.S. Food and Drug Administration to prevent abuses. Psychoactive drugs have been shown to be helpful in treating hyperkinesis, a restlessness that some experts believe derives from minimal brain damage or chemical imbalances. But what about the little boy in the Sarasota study who was thought to be hyperactive, and the many other children like him? If they get relief through drugs from stress caused by poor illumination and radiation, will that lead to later addiction to drugs or even alcohol?

Dr. Irving Geller, chairman of the Department of Experimental Pharmacology at Southwest Foundation for Research and Education in San Antonio, has found that abnormal conditions of light and darkness can affect the pineal gland, one of the master glands of the endocrine system. Experimenting with rats, Dr. Geller discovered that, under stress, they preferred water to alcohol. When left in continuous darkness over weekends, they went on alcoholic binges. And Nobel Prize winner Dr. Julius Axelrod had earlier found that the pineal gland produces more of the enzyme melatonin during dark periods. Injections of melatonin to rats on a regular light-dark cycle turned these rats into alcoholics.

That alcoholism may be related to the pineal gland is also under study by Dr. Kenneth Blum, a pharmacologist at the University of Texas Medical School. Under near-total darkness, rats with pineals drank more alcohol than water while rats *without* pineals drank more water than alcohol. When the animals were returned to equal periods of light and dark, rats with pineals retained their liking for alcohol. Applied to humans, Dr. Blum says, "It is possible that alcoholics may have highly active pineals."

Over the years, I have found that many biological responses are not necessarily responses to the total spectrum of light, but rather to narrow bands of wavelengths. When these are missing in an artificial light source, the biological receptor responds as in total darkness. For example, cool white fluorescent as well as ordinary incandescent bulbs lack the shorter wavelengths that we see as blue.

The hyperactive reaction to radiation from unshielded fluorescent tubes may have a correlation to the hyperactivity symptoms and severe learning disorders triggered by artificial food flavors and colorings, too. Dr. Ben F. Feingold of the Kaiser-Permanente Medical Center found that a diet eliminating all foods containing artificial flavors and colors brought about a dramatic improvement in fifteen of twenty-five hyperactive school children studied. Any infraction of the diet led within a matter of hours to a return of the hyperkinetic behavior.

This suggests the possibility of an interaction between wavelength absorption bands of these synthetic color pigments and the energy peaks and mercury vapor lines in fluorescent tubes. This means that, for example, two children in a family subjected to the same source of low level radiation would react differently if one preferred to drink cherry or strawberry pop and the other had a liking for some green- or yellow-colored soft drink. And this might mean that a reaction or allergy to fluorescent lighting could be eliminated two ways: by eliminating the absorbing material consumed when the child eats artificial color, or by eliminating the energy peaks in fluorescent tubes and other types of artificial lights.

At the molecular level, all chemicals and minerals have a maximum wavelength absorption band or wavelength resonance. Some responses are to wavelengths within the visible part of the total electromagnetic spectrum, but others are to longer or shorter wavelengths commonly referred to as general background radiation. For example, iron has a specific wavelength response. If a child won't eat spinach or some good source of iron, this child may not have the same reaction under the same radiation conditions as another who eats lots of spinach, raisins, etc.

Some drugs are known to make people more sensitive to sunlight, and the toxicity level of many drugs varies greatly depending on whether the drug is administered during the day or at night.

Jaundice in premature babies is now widely treated with blue light and psoriasis with long-wavelength ultraviolet (black light) after the patient is given a drug orally. It seems to me that if a particular ailment can be treated with certain wavelengths of light, living under an artificial light source lacking these wavelengths might logically contribute to causing the ailment in the first place. Conversely, long-term exposure to low-level or trace amounts of any radiation in excess of normal could produce abnormal responses or side effects over an extended period of time.

What all this means is that it now appears that there are biological responses to trace levels of radiation comparable to the equivalent trace levels in chemistry. Not long ago, one part per million was considered pretty insignificant and very difficult to measure accurately. But then it was discovered that one part per ten million, one part per billion, and one part per *trillion* can produce significant biological responses. Methods have been developed to measure these trace levels in chemistry, but as yet there are no methods to measure such low levels of radiation in light sources. We can only observe the reactions on various biological reactors such as bean plants, laboratory animals, and school children.

Lewis W. Mayron, Ph.D., at the Nuclear Medicine Research Laboratory of Veterans Administration Hospital, Hines, Illinois, and his wife, Ellen L. Mayron, a learning disabilities teacher at the Irene E. Hynes School, District 67, Morton Grove, Illinois, were the principal investigators in connection with the Sarasota school study. As this chapter goes to press, Dr. Mayron has several papers pending publication that discuss the possible direct relationships between low levels of radiation and hyperactivity in school children.

He points up an impressive list of references concerning the effects of electromagnetic radiation on animals and humans. Some of these effects include changes in electroencephalogram (EEG) frequency and amplitude in rabbits; subnormal EEG activity in a group of one hundred twenty people who had been exposed for more than one year to electromagnetic energy in the centimeter wavelengths; nervous exhaustion with irritability and, in some instances, abnormal slowness of the heartbeat; and increased incidence of reports of headache at the end of the workday as well as sleep disturbance and memory change.

Nikogosyan measured the blood cholinesterase activity in groups of rabbits exposed to varying levels of UHF energy for varying lengths of time. He found significant reductions in cholinesterase activity; only five to six weeks after the experiment ended did activity return to normal. The most pronounced changes were within fourteen days of the initiation of the experiment. It was also found that cholinesterase activity in the brain stems, hearts, and livers of the high-dose group of animals was low.

Noval et al. reported that rats exposed to extremely low-frequency non-ionizing radiations developed a twofold elevation of liver tryptophan pyrrolase activity (signifying stress) and a consistent depression in choline acetyl-transferase activity in the ventral portion of the brain that remains after removal of the cerebral hemispheres and the cerebellum. (Pons, medulla oblongata, and brain stem are left.)

Dr. Mayron also refers to the work of Gordon, who reported that UHF fields result in accumulation of acetylcholine along nerve fibers, perhaps explained by the above findings of reduced cholinesterase activity. Results from several investigators show that acetylcholine in small concentrations leads to cholinergic hyperactivity, whereas larger concentrations lead to a decrease in activity. This correlates with the behavior patterns exhibited by the rats radiated with UHF radiowaves and the rats in front of the television set; and this may also be the effect that is inhibited by placing a grounded aluminum screening on fluorescent fixtures in school rooms (as in the Sarasota study).

The amphetamine drugs given hyperactive children are basically stimulants. Why they act on hyperactive children to calm them is not fully understood. However, as increasing amounts of most stimulants, whether drugs or radiation, will increase activity up to a point (beyond which they develop a reversed reaction of lethargy, exhaustion, or stupor), are we simply overdosing these children with drugs in order to obtain a desired result? In querying a local pharmacist, I found that he advised that prescriptions of such drugs for children range as high as eighty to one hundred milligrams per day, whereas he thought twenty-five milligrams would keep a truck driver from falling asleep at the wheel.

Obviously a great deal more research is needed, and it will be years before all of the answers are known. By then, there will undoubtedly be many more new questions. However, one question of major importance right now is: In view of the alarmingly fast rise in learning disabilities and behavioral problems among school children, in crime, violence, terrorism, cancer—how long should the question of doing something about it be put off?

While the lighting and television industries generally have been publishing articles insisting that there is not as yet any conclusive proof that what affects bean plants and laboratory animals also affects man, two smaller manufacturing companies are doing something constructive. They have started to manufacture full-spectrum shielded fluorescent light fixtures that have already been specified in certain new construction by the Chicago school system, as well as by several other major school systems across the country. These two companies are Garcy Lighting, 1822 North Spalding Avenue, Chicago, Illinois 60647, and Forest Electric Company, located in Melrose Park, Illinois. Garcy is putting together the complete fixture, and Forest Electric is manufacturing a new type of full-spectrum fluorescent stabilizer to replace the present conventional ballast used in fluorescent fixtures.

This new type of stabilizer converts the AC line current to DC and thus eliminates the very objectionable sixty-cycle flicker that is recognized as contributing to headaches, eye strain, and fatigue, and which may be a factor in seizures in those subject to epilepsy. It virtually eliminates the heat produced by conventional AC ballasts, which is an added load on air-conditioning equipment. It also eliminates the usual AC humming noise. It produces a steady, continuous light and operates on approximately 19 percent less electric power, an important consideration in this day of energy crisis and increased power costs.

I worked with one of the fluorescent light manufacturers in developing an improved full-spectrum fluorescent tube. Basically, this was accomplished by adding black light (long-wavelength UV) to the blend of phosphors used in a fluorescent tube originally designed to closely duplicate only the visible wavelengths of natural sunlight for color-matching purposes. This was a very definite improvement in fluorescent lighting.

There was still the problem of mercury vapor lines, however. All fluorescent tubes do produce these very narrow but extremely intense spikes of energy in both the visible and ultraviolet wavelengths. The intensity of these mercury vapor lines varies in different types of fluorescent tubes—cool white, warm white, etc. This is a distinct disadvantage that must be weighed against all the advantages of fluorescent lighting.

(For some time I have been giving considerable thought to the problem of these mercury vapor lines in fluorescent tubes and of radiation from TV picture tubes, and I am exceedingly pleased to be able to report having received patents on two ideas that will completely eliminate these problems. They are still pretty much in the theoretical stage and will need considerable engineering development before they are operational, however.)

A recent encouraging development is that the Environmental Health and Light Research Institute will join with Roswell Park Memorial Institute in Buffalo, New York to form a new Center for Light Research and Studies. Roswell Park Memorial Institute, a leader in cancer research, is a part of the State University of New York at Buffalo, and it includes Roswell Park Hospital and Medical School. The new center will encourage, coordinate, and cooperate with research at other light research centers. This is the highest honor that EHLRI could have hoped for. No place has a finer reputation than Roswell Park.

The decision to invite EHLRI to become part of Roswell Park is a culmination of three years of experiments on the effects of different colors (wavelengths) of light on laboratory animals. Dr. Cora Saltareni, who directed the study, considers the results very significant.

EHLRI provided the lights and fixtures used in the experiments and will contribute all of its specialized equipment to the new light center. More than $250,000 worth of time-lapse cameras, controlling equipment, automatic dollies, and microscopic units are included. Regular seminars will be conducted to familiarize scientists with the latest methods and techniques in the field of light research.

The plan for the new center received the approval of Dr. Gerald P. Murphy, director of Roswell Park. At the present time, plans are being made to discuss funding for cancer research with Representative Paul Rogers (chairman of the House Subcommittee on Public Health and Environment). Such cancer research would include light as an important variable. A recent four-part series in *Eye, Ear, Nose and Throat* reported the findings of six research projects undertaken at major medical centers; these studied the effects of various wavelengths of light on tumor development.

The affiliation of EHLRI with a major university medical center is a milestone in the field of light research, which I have been concerned with since I began taking time-lapse pictures as a hobby in 1927.

Loyola University in Chicago, which awarded me an honorary Doctor of Science degree, was the first university interested in accepting the time-lapse equipment and incorporating light research into its department of biology. A site had already been picked out on campus when I was advised by the department chairman that research would be limited to the study of the effects of light on plants, and that no experiments on animals would be permitted. Since the effects of light on animals, and ultimately humans, had become the significant point of my interest, the plans with Loyola lapsed. (This was approximately twenty years ago.)

I then offered all the equipment and facilities to Michigan State, where I had been appointed a lecturer in the Department of Horticulture. The idea was voted down by the chairmen of all departments, however—there was no scientific substance to the idea of studying the effects of light on animals at that time. Offers to

212

several other universities were also declined. Only five years ago the University of Washington School of Fisheries turned down my son's Ph.D. research project on snails because it involved studying the influences of different kinds of light.

Light turned out to be *exactly* the problem several years later at the School of Public Health and Hygiene, Johns Hopkins University, Baltimore. I was giving a series of lectures there. Researchers were unable to propagate an organism that caused snail fever because of a small fluorescent desk lamp that was inadvertently being left on all night. When it was turned off with the rest of the lights, the problem was solved.

The tide of acceptance really turned in favor of EHLRI when experiments at four more major medical centers confirmed abnormal responses in bean plants and lab animals exposed to fluorescent lights. These experiments were carried on at the School of Public Health, Johns Hopkins; Nuclear Medical Research, Veterans Administration Hospital, Hines, Illinois; Department of Chemistry, University of South Florida; and Graduate School of the State University of New York, Buffalo.

We have finally learned that light is a nutrient much like food, and, like food, the wrong kind can make us ill and the right kind can help keep us well. Research has taken a giant step, but there is still much to be accomplished.

APPENDIX

PAPERS PRESENTED BY JOHN OTT

CHICAGO, ILLINOIS; American Association for the Advancement of Science, December, 1947. *"Time-Lapse Photography."*

WASHINGTON, D.C.; National Geographic Society, March, 1949. *"Flowers in Action."*

MADISON, WISCONSIN; The University of Wisconsin, May 9-10, 1956. "Engineering Photography."

MADISON, WISCONSIN; The University of Wisconsin, University Extension Division, May 2-25, 1957. *"Industrial Photography Applications."*

EAST LANSING, MICHIGAN; Michigan State University, Michigan Conference on Comparative Medicine, March 21-22, 1960. *"Effects of Light on Plants And Animals."*

NEW YORK, N.Y.; The New York Academy of Science, Conference on Photo-Neuro-Endocrine Effects in Circadian Systems, With Particular Reference to the Eye, June 6-8, 1963. *"Some Responses of Plants and Animals to Variations in Wavelengths of Light."*

OXFORD, ENGLAND; Fourth International Congress on Photobiology, July 26-30, 1964. *"Similar Photobiological Responses in Plants and Animals."*

TEL AVIV, ISRAEL; Federation Dentaire Internationale, July 10-17, 1966. *"Time-Lapse Photography."*

SEATTLE, WASHINGTON; The Washington Unit of the Academy of General Dentistry, May 13, 1967. *"Light Physiology."*

NEW YORK, N.Y.; The New York Academy of Dentistry, November 9, 1967. *"The Influence of Light on the Retinal Hypothalamic Endocrine System."*

HANOVER, N.H.; Dartmouth College; Fifth International Congress on Photobiology; August 26-31, 1968; *"The Effects of Environmental Lighting on Sexual Functions in White Rats."*

LAS VEGAS, NEVADA; American Association for Laboratory Animal Science, October 21-25, 1968; *"Environmental Effects of Laboratory Lighting."*

NEW YORK, N.Y.; Brooklyn Engineering Society; *"Engineered Light For Living."* 1968.

SPOKANE, WASHINGTON; Academy of General Dentistry, May 10, 1969. *"The Effect of Light on Life."*

SEATTLE, WASHINGTON; University of Washington; Eleventh International Botanical Congress, September 1, 1969. *"Influence of Ionizing Radiation on Directional Growth of Roots and Biological Clock Rhythms in Plants."*

NASSAU, BAHAMAS; National Eye Research Foundation, November 15, 1969, with Thomas G. Dickinson, M.D. *"Ocular Exposure to Trace Amounts of Ultraviolet Light as Adjunctive Therapy."*

JACKSONVILLE, FLORIDA; Southern Academy of Clinical Nutrition, April 3-5, 1970. *"Light."*

CHICAGO, ILLINOIS; Merchandise Mart; National Exposition of Contract Interior Furnishings, June 17, 1970. *"Lighting, Life and Health."*

NEW BRUNSWICK, N.J; American Association for Laboratory Animal Science, New York Metropolitan and Delaware Valley Branches, June 16-17, 1970. *"The Importance of Laboratory Lighting as an Experimental Variable."*

In addition to the above specific papers, John Ott has given more than one thousand lectures to civic organizations, garden clubs, university medical schools, hospitals and scientific organizations throughout the United States and in many foreign countries.

215

PUBLICATIONS BY JOHN OTT

TIME-LAPSE PHOTOGRAPHY. *The Chicago Naturalist,*
Vol. 10, pp 21-23, 1947.

STUDY OF THE DEATH OF IRRADIATED AND NON-
IRRADIATED CELLS BY CINEMICROGRAPHY (with
Robert Schrek, M.D.). *Archives of Pathology,* Vol. 53,
pp 363-378, 1952.

SEX LIFE OF THE PUMPKIN. *Atlantic,* Vol. 191, No. 2,
1952.

OBSERVATIONS OF EFFECTS OF LIGHT AND TEM-
PERATURE ON GROWTH OF FLOWERING PLANTS
AND FUNGI. *Bulletin of the Garden Club of America,*
Vol. 45, No. 4, July, 1957.

MY IVORY CELLAR. Chicago, Twentieth Century Press,
1958; Devin-Adair, Old Greenwich, 1971.

MEMORANDUM ON GROWTH RESPONSES OF
PLANTS AND ANIMALS TO VARIATIONS IN WAVE-
LENGTHS OF LIGHT ENERGY. *Tropicals,* Vol. VIII,
No. 3, p 22, January-February, 1964.

THE QUESTION OF LIGHT. *National Chinchilla Breeder,*
Vol. 20, No. 6, pp 17-18, June, 1964.

SOME OBSERVATIONS ON THE EFFECT OF LIGHT
ON THE PIGMENT EPITHELIAL CELLS OF THE
RETINA OF A RABBIT'S EYE. *Recent Progress in
Photobiology,* Blackwell, Oxford 1964.

SOME RESPONSES OF PLANTS AND ANIMALS TO
VARIATIONS IN WAVELENGTHS OF LIGHT ENERGY.
Annals, New York Academy of Sciences, Vol. 117,
Art. 1, pp 624-635, September 10,1964.

EFFECTS OF UNNATURAL LIGHT. *New Scientist,* Vol.
25, pp 294-296, February 4,1965. London.

EFFECTS OF WAVELENGTHS OF LIGHT ON PHYSIO-

LOGICAL FUNCTIONS IN PLANTS AND ANIMALS. *Illuminating Engineering*, Vol. LX, No. 4, Sec. 1, pp 254-261, April 1965.

A REVALUATION OF ULTRAVIOLET AS A VITAL PART OF THE TOTAL SPECTRUM. *Contacts* (Obrig Laboratories, Sarasota, Florida) April 1966.

THE INFLUENCE OF LIGHT. *Contacto, The International Contact Lens Journal*, Vol. 11, No. 1, pp 29-36, March, 1967.

E' PROPRIO INNOCUA LA TELEVISIONE? *L'Ingegnere*, Milan, November, 1967.

THE INFLUENCE OF LIGHT ON THE RETINAL HYPOTHALAMIC ENDOCRINE SYSTEM. *Annals of Dentistry*, Vol. XXVII, No. 1, March, 1968.

RESPONSES OF PSYCHOLOGICAL AND PHYSIOLOGICAL FUNCTIONS TO ENVIRONMENTAL LIGHT. *Journal of Learning Disabilities*, Vol. 1, No. 5, May, 1968.

RESPONSES OF PSYCHOLOGICAL AND PHYSIOLOGICAL FUNCTIONS TO ENVIRONMENTAL RADIATION STRESS. *Journal of Learning Disabilities*, Vol. 1, No. 6, June, 1968.

A RATIONAL ANALYSIS OF ULTRAVIOLET AS A VITAL PART OF THE LIGHT SPECTRUM INFLUENCING PHOTOBIOLOGICAL RESPONSES. *Optometric Weekly*, September 5, 1968.

ENGINEERED LIGHT FOR LIVING, *The Brooklyn Engineer*, Vol. XXII, No. 3, December, 1968.

A RATIONAL ANALYSIS OF ULTRAVIOLET AS A VITAL PART OF THE LIGHT SPECTRUM INFLUENCING PHOTOBIOLOGICAL RESPONSES. *Journal of the Maryland Optometric Association*, Vol. 1, No. 4, p 22, October-December, 1968.

INTERDISCIPLINARY COMMUNICATION NEEDS INCREASE. *Society of Photo-Optical Instrumentation Engineer's Journal*, Vol. 7, No. 2, January, 1969.

THE EFFECTS OF ARTIFICIAL LIGHT ON HEALTH. *Let's Live*, Vol. 37, No. 10, October, 1969.

THE IMPORTANCE OF LABORATORY LIGHTING AS AN EXPERIMENTAL VARIABLE. *[in press, Bucknell U.]*

REFERENCES AND SOURCES

BENOIT, J.; ASSENMACHER, I.: *The control of visible radiations of the gonadotropic activity of the duck hypophysis.* Recent Progress in Hormone Research, Vol. 15, pp 143-164. Academic Press, 1955.

BEVAN, B.M. and ZWEILLER, W.: *Ultraviolet Irradiation of Marine Aquaria.* Pre-publication, 1966.

BIRKNER, F.E.: *Photic influences on primate (Macaca mulatta) reproduction.* Laboratory Animal Care, Vol. 20, No. 2, Pt. 1, pp 181-185, April, 1970.

BISSONNETTE, T.H.: *Light and sexual cycles in starlings and ferrets.* Quarterly Review Biology, Vol. 8, pp 201-208, 1933.

BLUM, H.F.: *Carcinogenesis by Ultraviolet Light.* Princeton University Press, Princeton, N.J., 1959.

CHOW, K.L.; RIESEN, A. H.; NEWELL, F. W.: *Degeneration of retinal ganglion cells in infant chimpanzees reared in darkness.* Journal of Comparative Neurology, 17, pp 27-42, 1957.

CREMER, R.J.; PERRYMAN, P.W.; RICHARDS, D.H.: *Influence of light on the hyperbilirubinemia in infants.* Lancet, I :1227, 1958.

DANTSIG, N.M.; LAZAREV, D.N. and SOKOLOV, M.V. (USSR): *Ultraviolet Installations of Beneficial Action.* Washington, D.C., C.I.E., 1967.

DARREL, R.W. and BACHRACH, C.A.: *Pterygium Among Veterans.* Archives Ophthalmology, Vol. 70, pp 158-169, August, 1963.

de MAIRAN, M.: *Observation Botanique.* Paper Presented to the French Royal Academy of Sciences, Paris, 1729.

DUKE-ELDER, SIR W.S.: *Text-Book of Ophthalmology,* pages 826, 827, 828, St. Louis, C.V. Mosby, 1939.

DUNLOP, RICHARD: *Probing the Mysteries of Light.* Today's Health, Vol. 41, No. 3, pp 34-39, March, 1963.

EDITORIAL: *Light; A Double-edged Sword.* Annals of Ophthalmology, July, 1971.

EDITORIAL: *Light as Antidote to Stress.* New Scientist, No. 405, August, 1964.

EDITORIAL: "Medical News." Journal of The American Medical Association, Vol. 199, No. 5, pp 34, January 30, 1967.

EDITORIAL: *Phototherapy Exposure Tied to Retinal Damage.* Pediatric Currents, Vol. 20, No. 1, January, 1970.

ELLINGER, F.: *Medical Radiation Biology,* 44.3.0., Springfield, Illinois, Charles C. Thomas, 1957.

FELLER, R.P.; BURNEY, S.W.; SHARON, I.M.: *Some Effects of Light on the Golden Hamster.* IADR Abstracts, 1970. Paper presented to meeting of the International Association for Dental Research, New York, N.Y., March, 1970.

FREY, A.H.: *Behavioral Biophysics.* Psychological Bulletin, Vol. 63, No. 5, pp 322-337, May, 1965.

FRIEDMAN, A.H.; WALKER, C.A.: *Circadian Rhythms, II; Rat Mid-brain and Caudate Nucleus Biogenic Amine Levels.* Journal of Physiology, Vol. 197, pp. 77-85, 1968.

GELLER, IRVING: *Ethanol Preference in the Rat as a Function of Photoperiod.* Science, Vol. 173, pp 456-458, July 30, 1971.

HARDER, J.: *Acta Eruditorium Lipstae,* 1694 (reference listed by Wetterberg, Geller & Yuwiler, Science, p. 885, February 6, 1970.)

HATCH, A.; BALAZS, B.; WIBERG, G.S.; GRICE, H.C.: *Long-term Isolation Stress in Rats.* Science, Vol. 142, October, 1963.

HUGGINS, C.: *Endocrine Factors in Cancer.* Journal Urology, Vol. 68, p 875, December, 1952.

KORBEL, S.; THOMPSON, W.D.: *Behavioral Effects of*

220

Stimulation by UHF Radio Fields. Psychological Reports, Vol. 17, pp 595-602, 1965.

KREIG, W.J.S.: *The Hypothalamus of the Albino Rat.* Journal of Comparative Neurology, Vol. 55, No. 1, May, 1932.

LUCEY, J.F.: *Nursery Illumination as a Factor in Neonatal Hyperbilirubinemia.* Pediatrics, Vol. 4, No. 2, August, 1969.

MAYRON, LEWIS W.: *Environmental Pollution; Its Biological Effects and Impact on the Bioanalytical Laboratory.* American Association for the Advancement of Science. Meeting; paper presented, Chicago, December 27, 1970.

MOORE, R.Y.; HELLER, A.; BHATNAGER, R.K.; WURTMAN, R.J.; AXELROD, J.: *Central Control of the Pineal Gland: Visual Pathways.* Archives Neurology Vol. 18, pp 208-218, February, 1968.

MUMFORD, W.W.: Heat Stress Due to RF Radiation. Journal of Microwave Power, 4, 4, pp 242-254, 1969.

NOELL, W.K. and ALBRECHT, R.: *Irreversible Effects of Visible Light on the Retina: Role of Vitamin A.* Science, Vol. 172, pp. 76-80, April 2, 1971.

NOELL, W.K.; DELMELLE, M.C.; ALBRECHT, R.: *Vitamin A Deficiency Effect on Retina; Dependence on Light.* Science, Vol. 172, pp 72-75, April 2, 1971.

O'STEEN, W.K. and ANDERSON, K.B.: *Photically Evoked Responses in Visual System of Rats Exposed to Continuous Light.* Exper. Neurol. 30: pp 525-534, 1971.

ROWAN, W.: *Relationship of Light to Bird Migration and Developmental Changes.* Nature, Vol. 155, pp 494-495. 1925.

SCHREK, ROBERT: *Fashions in Cancer Research.* Year Book of Pathology and Clinical Pathology, 1959-1960.

SHIPLEY, R.: *Rod-cone Duplexity and the Autonomic*

Action of Light. Vision Research, Vol. 4, pp 155-177, May, 1964.

SISSON, THOMAS R.C.; GLAUSER, S.C.; GLAUSER, E.M.; TASMAN, W.; KUWABARA, T.: *Retinal Changes Produced by Phototherapy.* Pediatrics, No. 2, pp 221-227, August, 1970.

SPALDING, J.F.; HOLLAND, L.M.; TIETJEN, G.L.: *Influence of the Visible Color Spectrum on Activity in Mice.* Laboratory Animal Care, Vol. 19, No. 2, April, 1969.

STEPHENSON, N.R.: *A Sloping Screen Method for the Bioassay on Insulin in Mice.* Journal of Pharmacy & Pharmacology, Vol. 11, pp 659-665,1959.

TAKAYAMA, S. and OJIMA, Y.: *Photosensitizing Activity of Carcinogenic and Non-carcinogenic Polycyclic Hydrocarbons on Cultured Cells.* Japan Journal of Genetics, Vol. 44, pp 231-240, 1969.

TAYLOR, A. NEWMAN; WILSON, R.W.: *Electrophysiological Evidence for the Action of Light on the Pineal Gland in the Rat.* Experientia, Vol. 26, p 267, 1970.

THORINGTON, LUKE; PARASCANDOLA, L.; CUNNINGHAM, L.: *Visual and Biologic Aspects of an Artificial Sunlight Illuminant.* Journal of IES, Vol. 67, pp 33-41, October, 1971.

URBACH, F.; DAVIES, R.E.; FORBES, P.D.: *Ultraviolet Radiation and Skin Cancer in Man. Advances in Biology of Skin,* Vol. VII, Carcinogenesis, Chapter XII. Pergamon Press, 1966.

WALKER, C. A.; FRIEDMAN, A.H.: *Circadian Rhythms in the Toxicity of Cholinomimetics in Mice.* Fed. Proc., Vol. 27, No. 2, p 600,1968.

WEIHE, W.H.; SCHIDLOW, J.; STIMATTER, J.: *The Effect of Light Intensity on the Breeding and Development of Rats and Golden Hamsters.* International Journal Biometeor., Vol. 13, No. 1, pp 69-79, 1969.

WETTERBERG, L.; GELLER, E.; YUWILER, A.: *Harder-*

ian Gland; an Extraretinal Photoreceptor Influencing the Pineal Gland in Neonatal Rats? Science, Vol. 167, pp. 884-885, February 6,1970.

WURTMAN, R.J.: The Pineal and Endocrine Function. Hospital Practice, Vol. 4, No. 1, pp 32-37, January,1968.

WURTMAN, R.J.; WEISEL, J.: Environmental Lighting and Neuroendocrine Function; Relationship between Spectrum of Light Source and Gonadal Growth. Endocrinology, Vol. 85, No. 6, pp 1218-1221, December, 1969.

ZUCKERMAN, S.: Light and Living Matter. Transactions: Illuminating Engineering, Vol. 24, No. 3, 1959.

INDEX

224

225

Miller, Julian, 33
Mink, 120-21
Mitosis, 83
Monochromatic light, 82
Moon, 72
Moore, Robert Y., 90
Morning glories, 25-26, 29-30, 77, 123
Moscow, 65
My Ivory Cellar, 13, 91, 167, 175, 178, 195
Nader, Ralph, 141
Nagy, Rosemary, 81
Nat'l. Cancer Institute, 94
Nat'l Cen. Radiological Health, 143
Nat'l. Comm. Radiation Protection, 136
Nat'l. Eye Institute, 9
Nat'l Inst. Health, 76,162
Nat'l. Marine Fish Service Research Center, 186
Neoplastic disease, 189
Neurochemical channels, 89, 140
Neutral gray, 153
New College, 153
Newcomer, 109
Newell, F. W., 166
New England States Nat. Food Assoc., 187
N.Y. Academy Dentistry, 10
N.Y. Academy Sciences, 9, 77-78
N.Y. State University, 164
Night blooming flowers, 123-24
Night period (dark period) (see periodicity of light)
Nighttime radiation, 127
Niles, 168
Nishinomiya, Univ. of, 160
Nobel Prize, 176
Nocturnal animals, 128
Noell, 164
Northwestern Univ., 82, 178, 190
Northwood Mink Farms, 120, 155
Nutritional deficiency, 54-55

Obese, 99
Obrig Lab., 118-19
Oculo-endocrine system (see retinal-hypothalamic-endocrine)
Ojima, Y., 160

"On a Clear Day . . .", 147, 186
Ophthalmology, Annals of, 164
Ophthalmology, Survey of, 87
Ophthalmology, Textbook of, 109
Orbital tissues, 89
Oregon, Univ. of, 110
Orland, Frank J., 93
Oxygen, 106
Ozone lamp, 172

Pacific, University of, 157
Paramount Picture Corp., 147
Pediatric Currents, 163
Penn., State College, 134
Periodicity of light, 30, 43, 46, 61, 83, 90, 125, 156-57, 161-62, 167
Pfizer, Chas. & Co., 67, 72
Phase contrast microscope, 81, 165
Photobiology Congress, Fourth International, 10, 79-80
Photochemistry, 47
Photo-neuro-endocrine (see retinal-hypothalamic-endocrine)
Photophthalmia, 109
Photoreceptor, 38, 85, 89, 113, 139, 140
Photosyntheses, 8, 53, 85
Phytochrome, 127
Piglets, 162
Pigment epithelial cells, 80, 140, 146, 165
Pineal gland, 61, 85, 89,140,157-58, 189
Pistillate, 32
Pituitary gland, 13, 53, 60, 85, 90, 140, 176
Plastic greenhouse windows, 24-25, 55, 57, 61, 95, 153, 172
Poinsettias, 28-29
Polluted light, 154
Pop-eye, 116
Poultry, chickens, 53, 78
Presbytr.-St. Luke's Hosp., 93
Prime pelt season, 156
Primroses-dancing flowers, 23, 77
Prisms, 70
Production, increase in, 9
Protoplasm-streaming, 52, 79
Psychological Bulletin, 134

227

228

229